Workbook 6

William Collins' dream of knowledge for all began with the publication of his first book in 1819. A self-educated mill worker, he not only enriched millions of lives, but also founded a flourishing publishing house. Today, staying true to this spirit, Collins books are packed with inspiration, innovation and practical expertise. They place you at the centre of a world of possibility and give you exactly what you need to explore it.

Collins. Freedom to teach.

Published by Collins
An imprint of HarperCollins*Publishers* Ltd.
The News Building
1 London Bridge Street
London
SE1 9GF

HarperCollins*Publishers*
Macken House, 39/40 Mayor Street Upper
Dublin 1, D01 C9W8, Ireland

Browse the complete Collins catalogue at
www.collins.co.uk

© HarperCollins*Publishers* Limited 2021

10 9 8 7

ISBN: 978-0-00-836898-2

Second edition

Contributing authors: Karen Morrison, Tracey Baxter, Sunetra Berry, Pat Dower, Helen Harden, Pauline Hannigan, Anita Loughrey, Emily Miller, Jonathan Miller, Anne Pilling, Pete Robinson.

All rights reserved. No part of this publication may be reproduced, stored in a retrieval system, or transmitted in any form or by any means, electronic, mechanical, photocopying, recording or otherwise, without the prior written permission of the Publisher or a licence permitting restricted copying in the United Kingdom issued by the Copyright Licensing Agency Ltd, 5th Floor, Shackleton House, 4 Battle Bridge Lane, London SE1 2HX.

British Library Cataloguing in Publication Data
A Catalogue record for this publication is available from the British Library.

Commissioning editor: Joanna Ramsay
Product manager: Letitia Luff
Development editor: Karen Williams
Project manager: 2Hoots Publishing Services Ltd
Proofreader: Caroline Low
Cover designer: Gordon MacGilp
Cover illustrator: Ann Paganuzzi
Image researcher: Emily Hooton
Illustrators: Beehive Illustration (John Batten, Moreno Chiacchiera, Phil Garner, Kevin Hopgood, Tamara Joubert, Andrew Pagram, Simon Rumble, Jorge Santillan, Matt Ward)
Internal design and typesetting: Ken Vail Graphic Design Ltd
Production controller: Lyndsey Rogers
Printed in India by Multivista Global Pvt. Ltd.

With thanks to the following teachers and schools for reviewing materials in development:
Preeti Roychoudhury, Sharmila Majumdar and Sujata Ahuja, Calcutta International School; Hawar International School; Melissa Brobst, International School Budapest; Rafaella Alexandrou, Diana Dajani, Sophia Ashiotou and Adrienne Enotiadou, Pascal Primary School Lefkosia; Niki Tzorzis, Pascal Primary School Lemesos; Vijayalakshmi Chillarige, Manthan International School; Taman Rama Intercultural School.

Acknowledgements
The publishers wish to thank the following for permission to reproduce photographs.
Every effort has been made to trace copyright holders and to obtain their permission for the use of copyright materials. The publishers will gladly receive any information enabling them to rectify any error or omission at the first opportunity.

p1 Denis Ronin/Shutterstock, p79 Orla/Shutterstock.

Third-party websites, publications and resources referred to in this publication have not been endorsed by Cambridge Assessment International Education.

Contents

Topic 1 Healthy bodies
3D printed body parts — 1
What do you already know? — 2
Finding answers to scientific questions — 3
How does exercise affect us? — 4
What do you already know? — 7
Check yourself — 8
Preventing the spread of disease — 9
Keeping microorganisms out — 10
Circulatory system timeline — 11

Topic 2 Ecosystems
Draw and interpret a bar chart — 12
Match a graph to a data set — 14
Plot and interpret scatter graphs — 16
Find the food chains — 18
Food webs at school — 19
Investigating food safety — 20
Compile an action plan — 21
Using science to solve problems — 22
Investigating microplastics — 23

Topic 3 Materials
How do you plan and do an investigation? — 24
Planning a fair test — 25
How can we prove that gas has mass? — 27
Check your knowledge — 28
States of matter — 29
Heating curves — 30
Thermal conductivity — 31
Investigating electrical conductivity — 32
Materials used as insulators — 33
Describing changes — 34
Predicting changes — 35
How much will dissolve? — 36
Plot and analyse your results — 37
Applying what you know — 38
Observe a chemical reaction — 39
Did it react? — 40
Looking inside an egg — 41

Topic 4 Forces and energy
Force diagrams — 43
Make your own force meter — 44
Doubling and tripling the force — 45
Calculating the weight of different objects — 46
Calculating the weight on different planets — 47
Investigate forces in different directions — 48
Making sense of graphs — 49
Forces and their effects — 50
Designing an experiment — 51
Crash testing — 52
Looking back — 54

Topic 5 Light and electricity

All about seeing	55
How we see things	56
Building a periscope	57
Can light move around a corner?	58
Making a coin appear	59
Experiment with bending light	60
Using conventional symbols	61
Circuits and symbols	62
Using diagrams to build circuits	63
Testing series circuits	64
Changing components	65
Comparing circuits	66

Topic 6 Rocks and soil

Calculate percentages and graph them	67
Observing rock samples	68
Classifying rocks using a key	69
The rock cycle	70
Observe and compare your fossils	71
Which type of soil holds water best?	72
Which type of soil?	74

Topic 7 Earth in Space

Modelling the orbits of different planets	76
Comparing how fast the planets move	78
Pluto – a planet no more	79
Observing the Moon	81
Phases of the Moon	82

Appendices

Appendix 1: Electrical symbols	83
Appendix 2: Units for physical quantities	84

Topic **1** Healthy bodies

Student's Book p **2**
1.1 Modelling the human body

3D printed body parts

Modern technology has made it possible to print three-dimensional body parts. In some laboratories, scientists are even printing human organs.

Three-dimensional scans of the heart have been used to print this model.

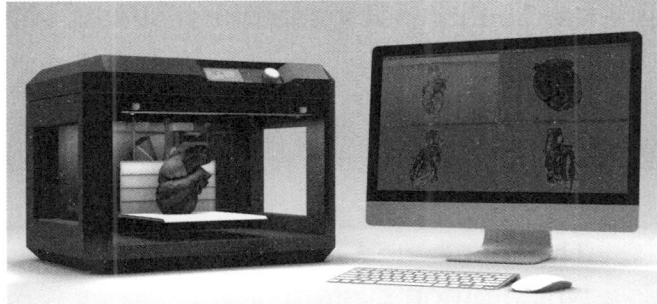

1 How could this technology be used by people who are teaching and training doctors?

_____

_____

2 How can 3D printed models help doctors diagnose problems with human organs?

_____

_____

3 Before an operation, doctors can scan an organ and print a model of it. How can this make the operation safer?

_____

_____

4 Prosthetics are artificial body parts used to replace limbs (for example, legs) when people have lost or damaged their own limbs. How could 3D printed prosthetics be an advantage:

  a in rural areas where there are no specialists to make prosthetics?

  _____

  _____

  b for children who outgrow their prosthetic limb?

  _____

  _____

Topic 1 Healthy bodies

Student's Book p 4
1.2 Organs and organ systems

What do you already know?

Answer as many of these questions as you can.

1 The lungs, nose and windpipe are part of which organ system?	
2 The mouth and stomach are part of which organ system?	
3 What happens to food once it leaves your stomach?	
4 Which organ pumps blood around the body?	
5 Which organ system has tendons and ligaments?	
6 What is the job of a ligament?	
7 What are the jobs of the ribs?	
8 Describe how the blood and heart work together.	
9 What is the main function of the lungs?	

Topic **1** Healthy bodies

Student's Book p 4
1.2 Organs and organ systems

Finding answers to scientific questions

There are different methods of answering scientific questions, including:

- Observe and measure
- Carry out a survey
- Carry out a fair test
- Look for information in books, on the internet or ask an expert

Which method would you use to find the answers to these questions about body systems?

A Where is the heart situated?
B In what way does exercise affect the heart rate?
C What happens to your breathing rate when you exercise?
D Why is saliva important for digestion?
E What does the inside of a human lung look like?
F Is there a difference between the pulse rate of boys and girls?
G Do older people take more breaths per minute than younger people?
H How many people know their blood type?
I Which body organs can be successfully transplanted?
J What is the difference between a vein and an artery?
K Do athletes have lower resting pulse rates than people who are unfit?
L Why is exercise important for a healthy body?
M What proportion of students in our school have asthma?

Write each letter in the correct column in the table below.

Observe and measure	Use sources to find information	Carry out a survey	Carry out a fair test

Topic **1** Healthy bodies

How does exercise affect us?

Student's Book p 6
1.3 The circulatory system

Plan a fair test investigation to find out what happens to your pulse rate after three different types of exercise. Identify the independent, dependent and control variables.

1 Planning

The dependent variable is _____

_____

The independent variable is _____

_____

The control variable is _____

2 Prediction

I think that when I exercise _____

_____

3 Describe how you will carry out the investigation.
Think about what you will measure and how you will make it a fair test.

_____

_____

_____

4 In the space below, add comments on your plan from another pair of students.

_____

_____

_____

4

continued

Topic **1** Healthy bodies

5 Results:

Type of exercise	Pulse rate
Resting	

6 Show your results as a bar graph.

continued

Topic 1 **Healthy bodies**

7 Conclusion (what we found out): _____

_____

_____

I think this is because _____

_____

_____

8 Look at your prediction.
Does the evidence support your prediction? Explain.

_____

_____

_____

_____

9 What could you do to improve your investigation?

_____

_____

_____

_____

Topic 1 Healthy bodies

Student's Book p 8
1.4 The respiratory system

What do you already know?

I predict that, after running on the spot for one minute, _____

_____

Method

1 _____

2 _____

3 _____

Results

	Breaths in one minute while at rest	Breaths in one minute after one minute of exercise
My result		
My partner's result		

Observations

Was my prediction correct? _____

Explanation of my results: _____

_____

How reliable is my data? Do I need to repeat any of the measurements? When do people breathe more often and why? Link your answer to your results from investigating the effects of exercise on heartbeat.

_____

_____

Check yourself

1 What does 'infectious disease' mean? Tick the true statements.

☐ a disease that spreads from one organism to another

☐ a disease that does not spread between organisms

☐ a disease with many symptoms

☐ a disease of the skin

2 Read the information about different diseases and their symptoms.

  a Fill in the type of microorganism that causes each disease.

  b Add an example of your own to the table.

Disease	Symptoms	What causes it
athlete's foot	red and itchy skin between the toes	
chicken pox	fever, raised red spots with yellow tops	
cholera	vomiting, very bad diarrhoea, muscle cramps	
colds and influenza (flu)	fever, sore throat, aches	
tuberculosis (TB)	fever, coughing up blood	

3 Give an example of a disease spread by animals.

4 Explain why infectious diseases can easily be spread in a confined space such as a bus.

Topic **1** Healthy bodies

Student's Book p **12**
1.6 Preventing the spread of disease

Preventing the spread of disease

1 Shazia has the flu.

  a Describe some of the symptoms she might have.

  _____

  b Describe two ways Shazia could prevent spreading the disease to the people she lives with.

  _____

  _____

2 The World Health Organization (WHO) keeps track of the number of cholera cases that are reported each year. This graph shows this information for the years 2000 to 2016.

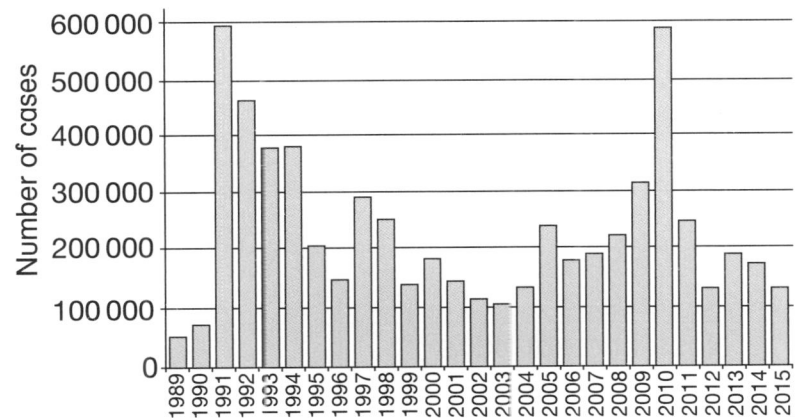

  a Describe the pattern shown on the graph.

  _____

  b In 2010 a massive earthquake struck Haiti and over 3 million people were left without access to basic services. Nine months after the earthquake a cholera epidemic broke out. Use the data on the graph to estimate how many people were affected.

  _____

  c What happened to the number of cholera cases worldwide after this epidemic was brought under control in 2012?

  _____

  _____

Topic 1 Healthy bodies

Student's Book p 14
1.7 Defence mechanisms

Keeping microorganisms out

1 Complete this table to summarise what you learned about your body's defence mechanisms in this unit.

Defence mechanism	Physical or chemical barrier?	What it does
Intact skin		
Broken skin (wound or cut)		
Stomach acid		
Tears and saliva		
Nose and ear hairs		
Tiny hairs in your windpipe		

2 When you are sick with a cold, you often have a very runny nose and you cough up lots of phlegm and mucus. Why do you think this happens?

_____

_____

_____

Topic 1 — Healthy bodies

Student's Book p **16**
1.8 History of science: Human anatomy

Circulatory system timeline

Complete the timeline below to summarise how knowledge and understanding of the human circulatory system has changed over time.

Ancient Greeks
(500–300 BCE)

> They believed

Ancient Romans
(Galen 129–c.210 CE)

> Galen wrote that

Arab scientists
(950–1300 CE)

> Studies in centres like Baghdad showed that

European scientists
1300–1500

> Vesalius proved that

1600–

> William Harvey built on earlier work and suggested that
>
> Malpighi discovered veins and arteries were connected by capillaries using a microscope.

> Today it is possible to …

11

Topic 2 Ecosystems

Draw and interpret a bar chart

Student's Book p 20
2.1 Working with graphs

The table below gives the mass of plastic found on the surface of different parts of the world's oceans.

Region	North Pacific	South Pacific	Indian Ocean	North Atlantic	South Atlantic	Mediterranean
Mass of plastic (tonnes)	96 400	21 000	59 130	56 470	12 780	23 150

1 Draw a bar chart using the axes provided to show this data. Give your bar chart a suitable heading and label the axes clearly.

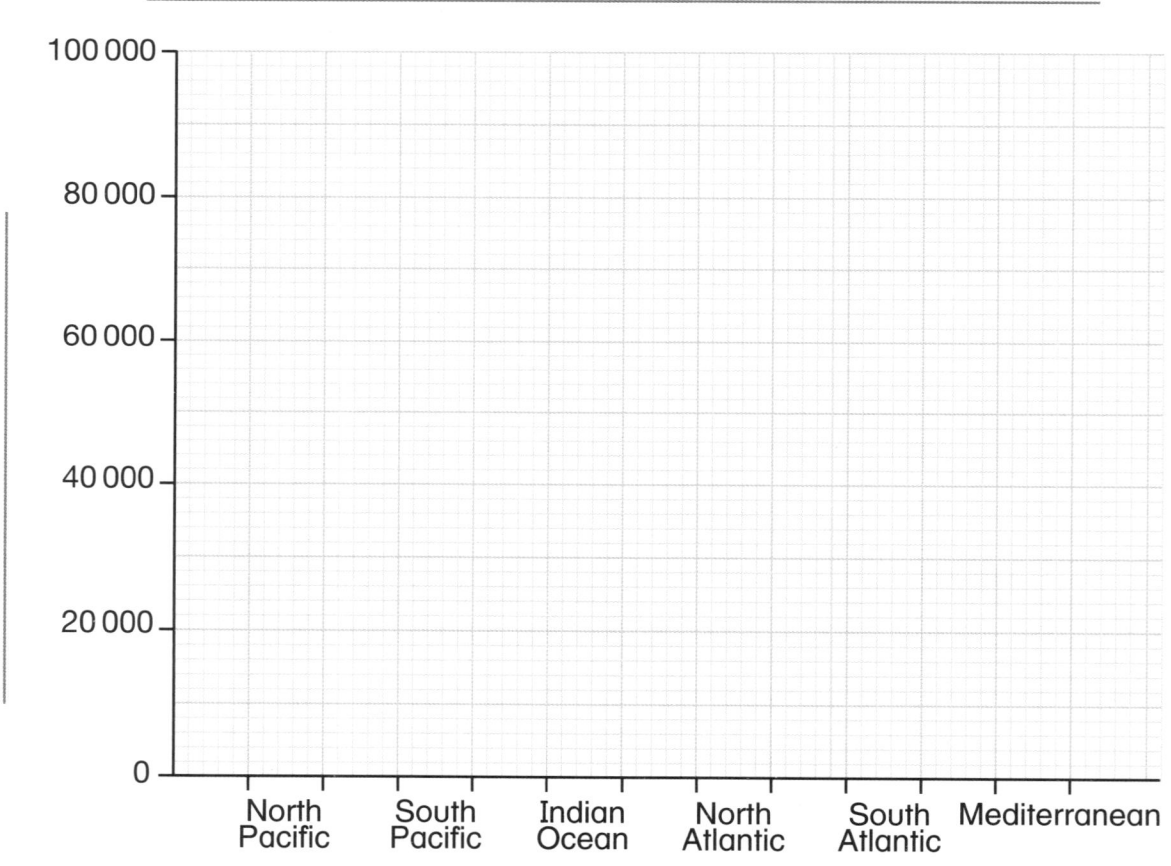

continued

2 Which region has the most plastic pollution?

_____

3 In which region would you expect to find the fewest people living near the coast? Why?

_____

4 What can you conclude from this graph?

_____

_____

_____

5 Which ocean region/s are closest to your country? Write a sentence describing how plastic pollution in that region compares with the rest of the world's oceans.

_____

_____

_____

_____

Topic 2 Ecosystems

Match a graph to a data set

Student's Book p 20
2.1 Working with graphs

A group of university students tracked the amount of an insecticide called DDT in the bodies of fish in a local lake over a period of ten years. For the first three years, the amount of DDT increased steadily, but after that it dropped slowly year by year.

1 Tick the graph that best matches this information.

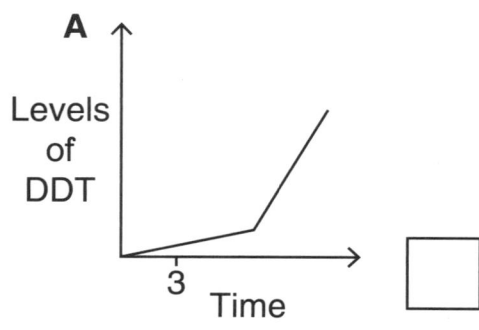

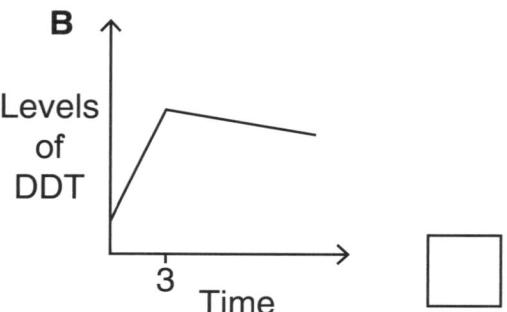

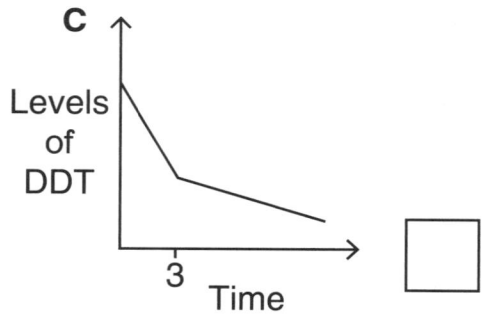

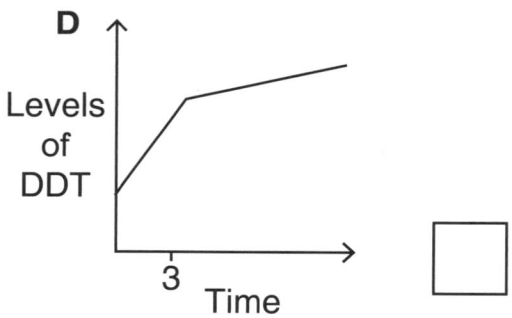

2 Could you show this data effectively on a bar chart? Explain your answer.

_____

_____

3 Could you use a scatter graph to show this data effectively? Explain your answer.

_____

_____

continued

Topic 2 Ecosystems

In the USA, the use of DDT as an insecticide was banned in 1972. In southern Africa, DDT was widely used to kill malarial mosquitoes until it was banned in 1996.

4 What do you think happened to the amount of DDT found in animal's bodies in the USA in 1973? Why?

_____

_____

_____

5 Do you think banning the use of DDT in southern Africa in 1996 necessarily caused an increase in the cases of malaria? Why?

_____

_____

_____

Topic 2 Ecosystems

Plot and interpret scatter graphs

Student's Book p 20
2.1 Working with graphs

A scientist measured the amount of coral growth on coral reefs (in millimetres per year) at twelve sites. She also recorded the mean surface temperature at each site. She concluded that higher temperatures were linked to a smaller amount of coral growth.

1 Plot twelve points on this scatter graph that support the scientist's conclusion.

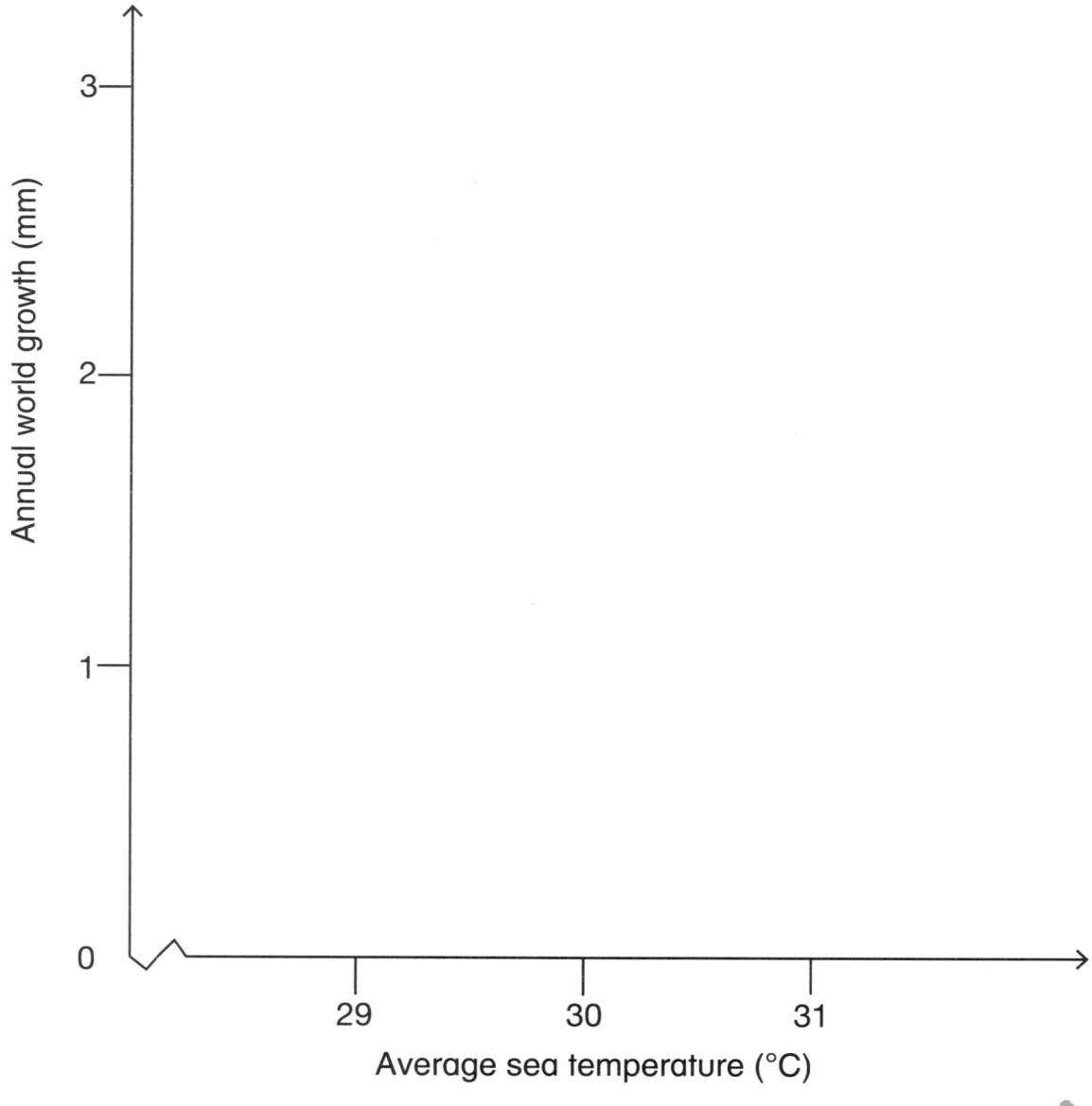

continued

Topic 2 — Ecosystems

A student kept track of the number of mobile phone data cards sold at a kiosk and the number of times the life guards had to help people in the water each week during the summer holidays. He drew this graph to show his data.

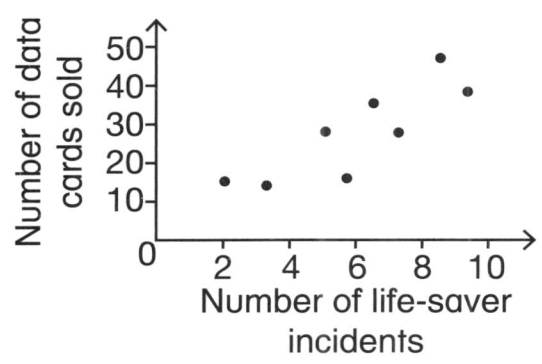

He wrote this about the graph:

The more life-saver incidents there are, the more data cards are sold. So, it seems like life-saver incidents cause people to buy data.

2 Is the first sentence correct?

_____

3 Is it reasonable to conclude that life-saver events cause data sales? Explain your thinking.

_____

_____

_____

Topic 2 Ecosystems

Student's Book p 22
2.2 Food chains and food webs

Find the food chains

Draw three different food chains for each habitat.

Habitat A – a tropical sea

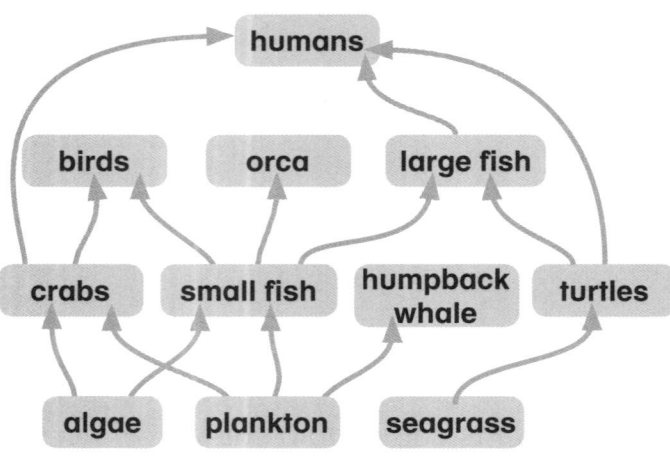

Habitat B – a freshwater pond

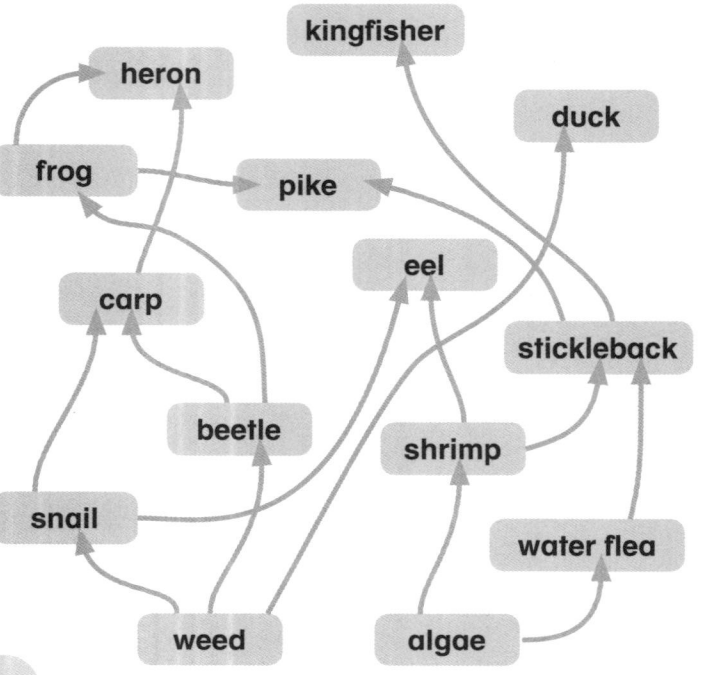

Topic **2** Ecosystems

Student's Book p **24**

2.3 Energy transfers in food chains

Food webs at school

A group of students chose a rose bush in the school garden as an example of a habitat. These are the questions they asked about it:

- Did you see those little green flies on the rose buds?
- I saw a ladybird beetle. What do you think it was doing on the rose bush?
- I wonder what ladybirds eat?
- Does anything eat ladybirds? They don't look very tasty.

1 What could you do to find out the answers to these questions?

_____

_____

2 One of the students found this information:

> Greenflies, or aphids, have sharp tube-like mouth parts. They use their mouth parts to make a hole in a plant and then they suck sweet liquid from the plant. Ants carry aphids from plant to plant. Insects such as ladybird beetles feed on the aphids. Some spiders and beetles, as well as swallows and swifts, will eat ladybirds, but most other birds will not because they taste disgusting. Geckos sometimes eat ants.

Use the information to draw a food web for the rose bush habitat in the space opposite.

3 Choose one habitat in your own school environment. Make a poster to show a food chain for the habitat. Label the producers and consumers and write short notes to explain how energy is transferred in this food chain.

19

Investigating food safety

Student's Book p 26
2.4 Toxins in food chains

The European Food Safety Authority (EFSA) monitors and tests foods for contaminants.

This graph shows the most common types of foods that tested positive for contaminants in 2015.

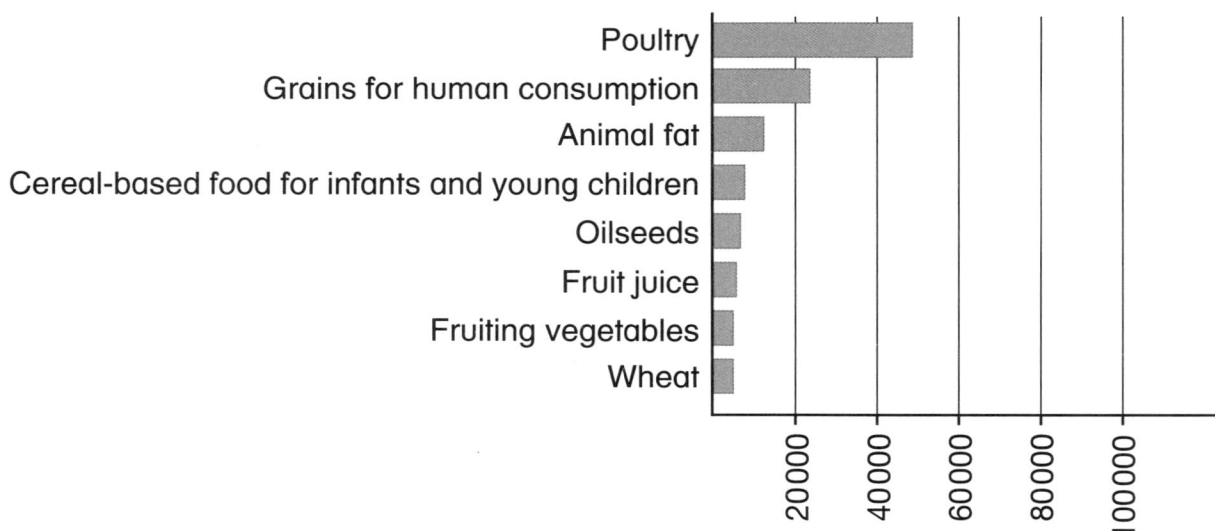

Choose three types of food shown on the graph.

Complete the table for each food type.

Type of food			
Number of contaminated samples			
How contaminants may have entered the food chain to affect this type of food			
Possible effects on people, animals and the environment			

Topic 2 Ecosystems

Student's Book p 26
2.4 Toxins in food chains

Compile an action plan

| An action plan shows what you are going to do about something and how you are going to do it. | An action plan should include:
• a sketch map of the problem area
• information about why the area is important
• what problems you have found
• how the area could be improved
• some practical suggestions for improving the area. |

1 Work in a group to identify one area where you think toxins might enter food chains.

Description of area:

_____

_____

_____

_____

Sketch map of the area:

2 Why is this area important?

_____

_____

3 What are the most important problems found in the area?

Problem	Possible cause	Effects

4 What are the suggested next steps to reduce or solve the problems?

_____

_____

21

Topic 2 Ecosystems

Using science to solve problems

Student's Book p 28
2.5 Plastics

Can plants and bacteria solve our plastic problems?

Scientists are working to design and develop more environmentally friendly materials to replace oil-based plastics.

One of the problems with plastic is the amount of waste. Bioplastics help to reduce plastic waste, but producing these materials in large amounts costs lots of money and special facilities are needed to dispose of them properly.

Plastics have been made from plants for many years. Scientists take sugars from plants like sugar cane and use microorganisms to convert these to the raw materials that can be formed in to plastics. The challenge is to produce plant-based plastics that break down in the environment but which are durable enough to use – if they break down while you are using them, then they aren't very useful!

Many bioplastics break down within three months in controlled conditions to form compost for gardens and crops. However, when these plastics end up buried in dumps or landfills, they don't break down quickly because the conditions aren't right. This is one of the challenges scientists need to address.

Using plants to make plastics is more sustainable than using oil, but we still need land and water to grow the plants. Some researchers are developing plastics using cyanobacteria. These bacteria use sunlight and carbon dioxide (by photosynthesis) to create the sugars needed to make the bioplastic. This reduces the need for land and water resources.

Bioplastics need to be cheap, lightweight, durable enough to use and not harm the environment.

1   What are bioplastics?

2   How can bioplastics help to reduce plastic waste problems?

3   What raw materials can be used to make bioplastics? What are the advantages of each?

4   Identify two challenges raised in the article and suggest how these could be overcome.

Topic 2 Ecosystems

Student's Book p 28
2.5 Plastics

Investigating microplastics

Read the article on PCM B15 about the work done by Australian scientists.

Complete this table to summarise what the scientists did and what they learned.

What scientific question did they ask?	
What type of scientific enquiry did they choose to try to answer the question? Why?	
Briefly describe what they did.	
How did they collect and record observations?	
What did their results show?	
What conclusions did they reach?	
How do you think their investigation could be improved?	
What did you find most interesting about their work? Why?	

Topic **3** Materials

Student's Book p **32**
3.1 Scientific enquiry

How do you plan and do an investigation?

The possible steps in an investigation are listed in the first column of the table.

1 Complete the second column by describing what you would do in each step.
2 Give an example to help explain what each step means in the third column.

Step	Description	Example
Ask a question		
Make a hypothesis		
Predict the outcome		
Plan a test		
Carry out your test		
Collect and record evidence		
Analyse your results		
Draw a conclusion		

Topic **3** Materials

Student's Book p **32**
3.1 Scientific enquiry

Planning a fair test

Plan a fair test to answer this question:

Does a large mug of boiling water cool at the same rate as a small mug of boiling water?

1 Planning

    a Predict what you think the outcome will be.

    _____

    b What variables will you change?

    _____

    c What variables will you keep the same?

    _____

    d What risks are there in this investigation?

    _____

    e How will you minimise these?

    _____

2 Carrying out the test

    a What equipment will you need?

    _____

    b What observations and measurements will you take?

    _____

    c How will you make sure your observations and measurements are accurate?

    _____

    d How will you record your observations/measurements?

    _____

continued

Topic 3 Materials

3 Analyse, evaluate and draw conclusions

  a Draw a suitable graph to show your results.

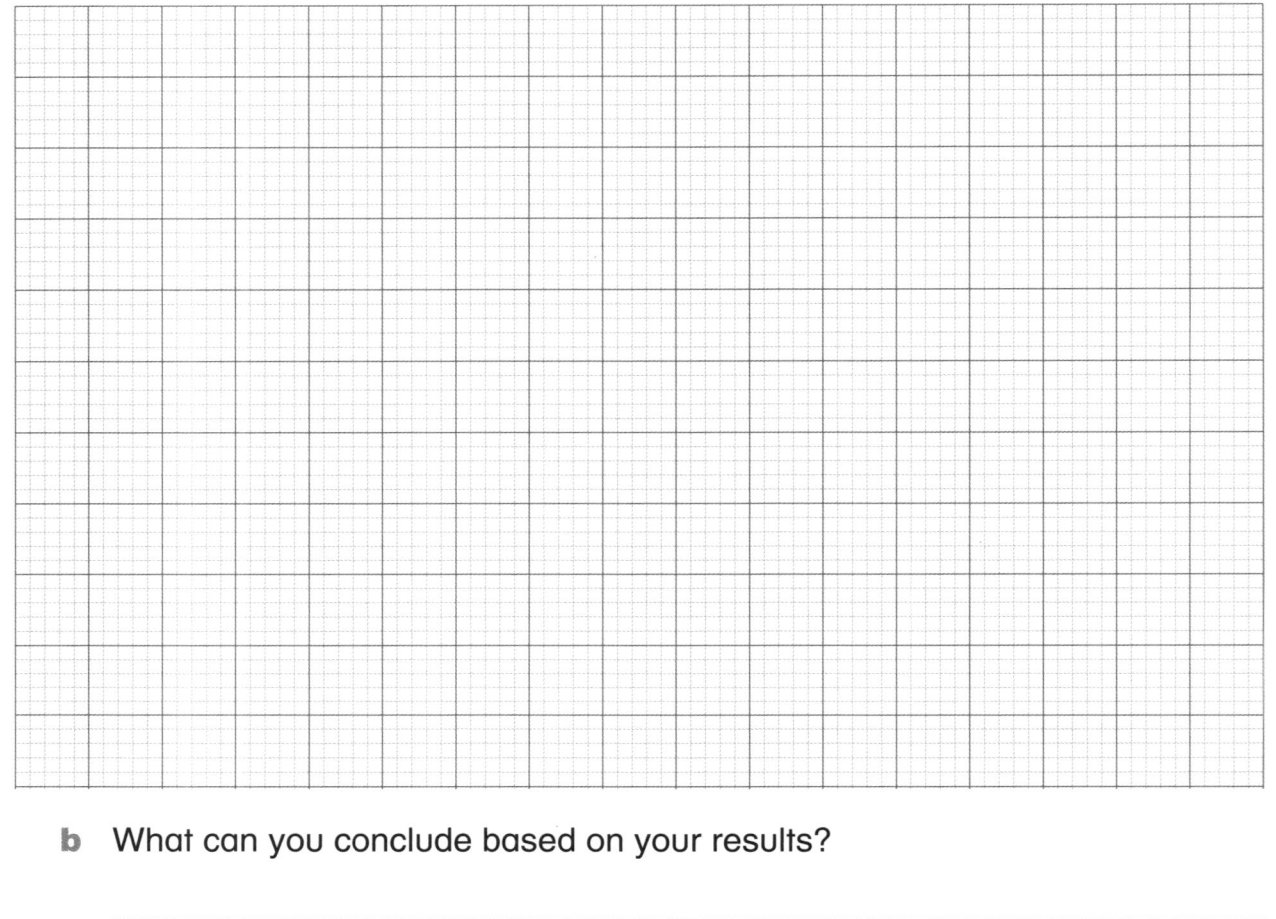

  b What can you conclude based on your results?

  _____

  _____

  c Look back at your prediction. Does the evidence support it? Explain.

  _____

  _____

  d Suggest one thing that you could do to improve this investigation.

  _____

  _____

Topic 3 Materials

Student's Book p 34
3.2 Solids, liquids and gases

How can we prove that gas has mass?

Our question:

Our hypothesis:

Our prediction:

Equipment:

Method:

Observations and results:

Conclusion:

Topic 3 Materials

Student's Book p 34
3.2 Solids, liquids and gases

Check your knowledge

1 Complete the table by adding ticks to show what state each substance would be in at a room temperature of 20 °C.

Substance	Solid	Liquid	Gas
propane			
milk			
helium			
aluminium			
oil			

2 Label the particle diagrams and state one property of a substance in each state.

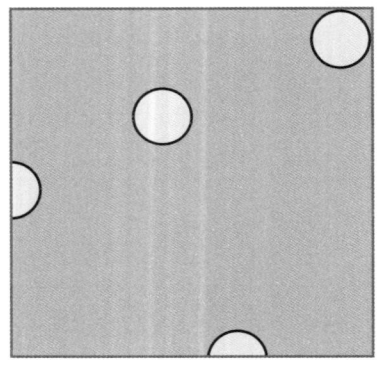 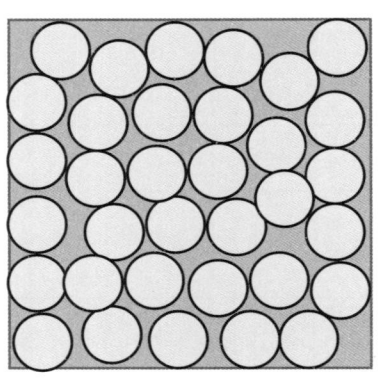 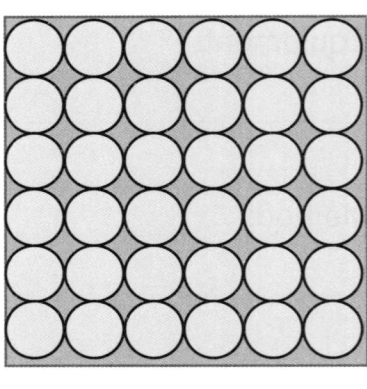

_____   _____   _____

_____   _____   _____

3 Use the particle model to explain why a balloon that you have blown up will get smaller if you put it in the fridge.

_____

_____

_____

States of matter

Topic 3 Materials

Student's Book p 36
3.3 Heat and changes of state

Add labels to the diagram to summarise what you have learned about changes of state.

Choose words from the box.

> solid   liquid   gas   gain heat   lose heat   melting
> evaporation   condensation   boiling   freezing

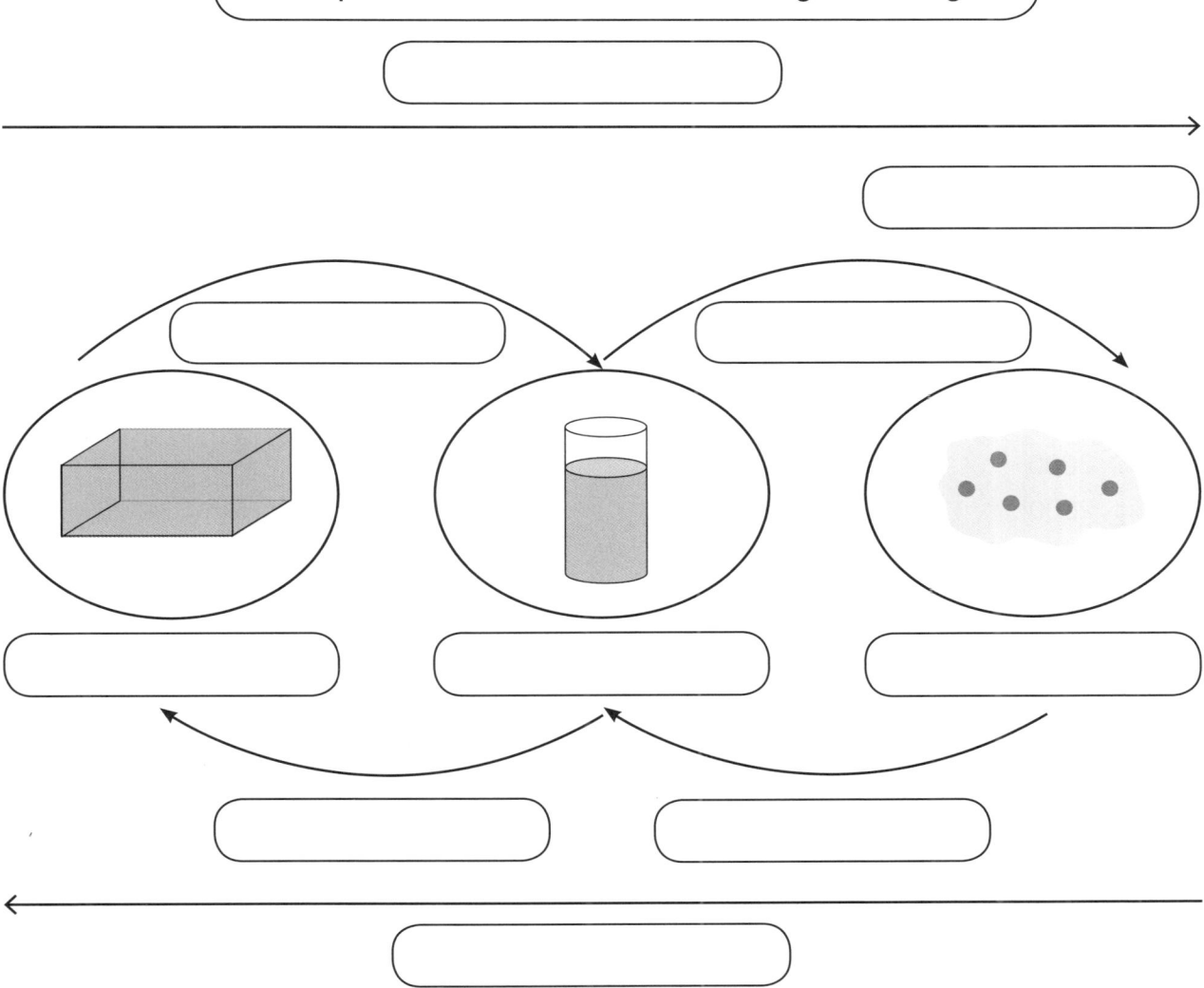

Topic 3 Materials

Student's Book p 36
3.3 Heat and changes of state

Heating curves

When a substance is heated at a fixed rate, scientists measure its temperature at intervals.

This graph shows the general structure of a heating curve.

1 Study the graph and discuss what it shows.

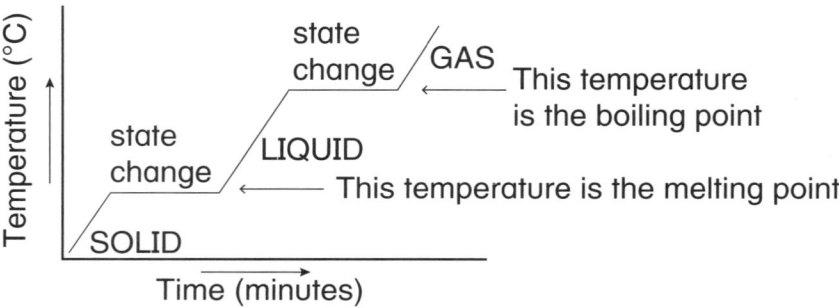

2 This is the heating curve for iron.

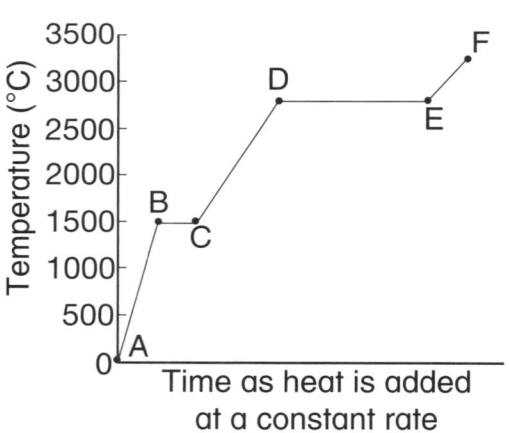

a Identify the approximate melting point of iron. _____

b At what temperature does iron change from liquid to gas? _____

c In what state would iron be at 2450 °C? _____

d Compare the length of the horizontal lines on the graph. What does the difference suggest?

_____

Topic **3** Materials

Student's Book p 38
3.4 Conductivity

Thermal conductivity

1 Marc and Jabu decided to compare the thermal conductivity of iron, copper and brass.
They set up this experiment.

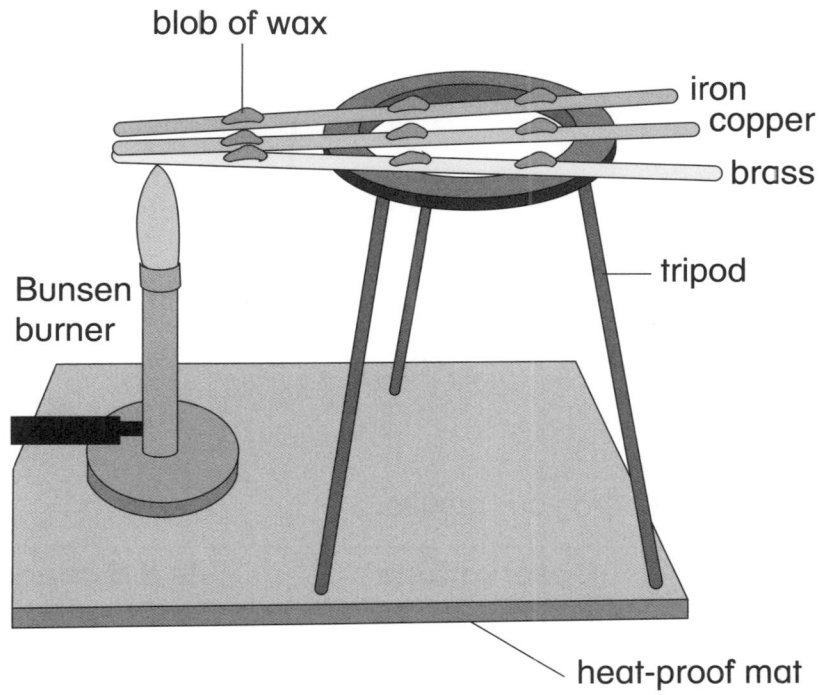

a How will this experiment help them rank the metals in order from best to worst conductor of heat?

_____

_____

b List two safety precautions for this experiment.

_____

_____

c The wax on the copper melts first, then on the brass and then on the iron. What does this tell you about the thermal conductivity of these metals?

_____

_____

Topic 3 Materials

Investigating electrical conductivity

1 Test various materials to see whether they conduct electrical current.

Student's Book p 38
3.4 Conductivity

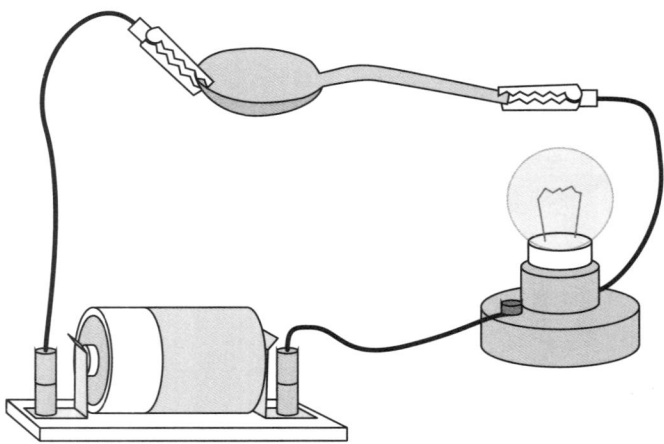

2 Complete this table to record your results.

Material tested	Observations	Is it a conductor?

3 Write a conclusion to your investigation to make a general statement about which materials conduct electricity.

Conclusion:

_____

_____

_____

Topic 3 Materials

Student's Book p 40
3.5 Conductors and insulators

Materials used as insulators

1. Use this table to record the number of insulators made from each material per appliance.

Appliances	Materials				
	Plastic	Rubber	Glass	Wood	Other
Total					

2. Combine your results with those of another pair. Write the totals below.

Plastic:	Rubber:	Glass:	Wood:	Other:

3. Draw a suitable graph to display your combined results.

4. Comment on what your graph shows.

_____

_____

Topic 3 Materials

Student's Book p 42
3.6 Reversible physical changes

Describing changes

1. Describe the changes in the pictures. Try to use scientific words in your descriptions.
2. Say what has caused each change.
3. Circle the changes that are reversible.

Topic **3** Materials

Student's Book p **42**
3.6 Reversible physical changes

Predicting changes

1 Describe what happens to each substance when it is changed by the process in column 2. Predict whether each change is reversible.

Substance	Process	Description of what happens	Is the change reversible?
Water	Freezing		
Chocolate	Melting		
Dough	Baking		
Wood	Burning		
Water	Boiling		
Salt and water	Mixing		
Metal	Magnetising		
Candle wax	Melting		
Clay	Firing in a kiln		
Plastic	Melting		
Porridge	Cooking		
Egg	Boiling		
Sand and water	Mixing		
Paper	Soaking in water		

2 What patterns can you see in the table?

_____

_____

3 What can you conclude about reversible changes from this information?

_____

_____

_____

35

How much will dissolve?

Student's Book p 44
3.7 Temperature and dissolving

You have planned your fair test investigation as a group on a large sheet of paper.

1 What is your prediction? Why? _____

_____

2 Record your results below. Design a table.
Remember that you want reliable results.

3 Do your results agree with your prediction? _____
Suggest reasons for any difference.

_____

Topic **3** Materials

Student's Book p 44
3.7 Temperature and dissolving

Plot and analyse your results

1. Look at your table of results on page 36. Highlight any results that don't seem to fit the pattern. Ignore these when you draw your graph.

2. Draw a suitable graph to show your results.

3. What can you conclude from this data?

   _____

   _____

   _____

4. If you did this investigation again, what could you do to improve it?

   _____

   _____

   _____

Topic 3 Materials

Applying what you know

Student's Book p 44
3.7 Temperature and dissolving

1 What happens to sugar when you put it into water?

_____

2 Explain why dissolving is a reversible change.

_____

3 Add the words *increases* or *decreases* to make the sentences true.

As the temperature of water _____, the speed at which sugar dissolves increases.

As the temperature of water _____, the amount of sugar that can be dissolved decreases.

4 Look at this graph from a student investigation.

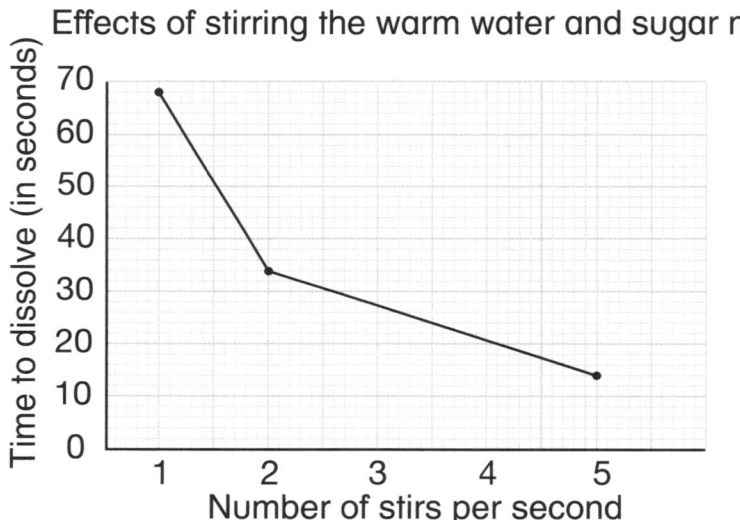

Use the graph to suggest one other way of speeding up dissolving.

_____

5 Use what you know about the particle model to explain why:

   a sugar dissolves faster in warmer water

   _____

   b more sugar can dissolve in 100 ml of hot water than in 100 m of cold water

   _____

   c stirring the solution makes the sugar dissolve faster.

   _____

Topic **3** Materials

Student's Book p **46**
3.8 Chemical reactions

Observe a chemical reaction

1 Describe the appearance of the reactants.

Vinegar: _____

_____

Bicarbonate of soda: _____

You will need:
- vinegar
- bicarbonate of soda
- a beaker
- a spoon

_____

2 Pour a small amount of vinegar into the beaker. Add a teaspoon of bicarbonate of soda.

3 Describe what happens.

_____

_____

4 Describe the appearance of the vinegar and bicarbonate of soda after the reaction has taken place.

_____

_____

5 How do you know that this change is irreversible?

_____

_____

6 Students sometimes use this reaction to make a model of red lava erupting from a volcano. How do you think they do this?

_____

_____

Topic 3 Materials

Did it react?

Student's Book p 48
3.9 Identifying chemical reactions

1 Complete the table to classify the changes.

Change	Physical or chemical?	Reason
A plate is dropped and it breaks		
A green banana ripens and turns yellow		
A firework explodes		
Hot chocolate powder dissolves in water		
A water purifying tablet starts to fizz in water		

2 What is the one thing that always happens in a chemical reaction?

_____

3 Describe what is happening in each diagram to show that a chemical reaction is taking place. Circle the diagram that shows unsafe behaviour.

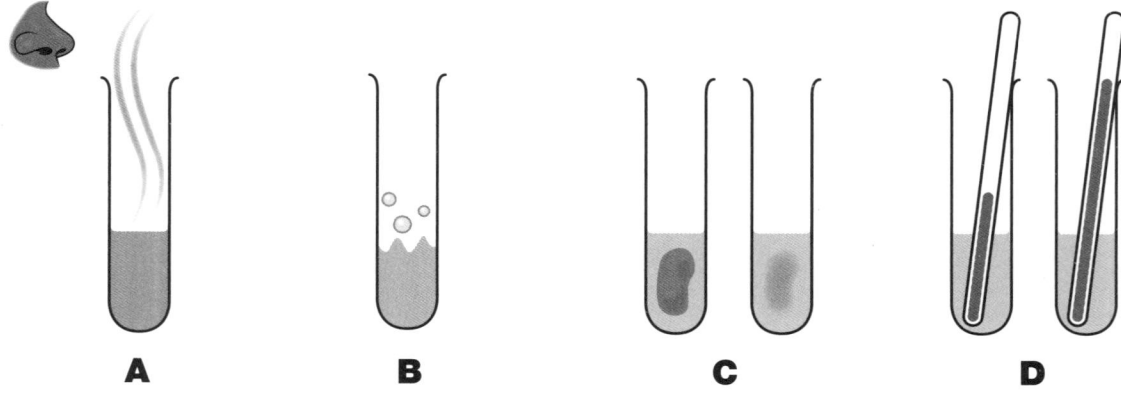

A _____

B _____

C _____

D _____

Topic 3 Materials

Student's Book p 48
3.9 Identifying chemical reactions

Looking inside an egg

Can you use a chemical reaction to look inside an egg?

You will need:
- an egg
- white vinegar
- a beaker
- a spoon
- a bowl of clean water

1. Describe the colour and feel of the egg shell.

   _____

2. Carefully put the egg into the beaker. Make sure you do not crack the shell.

   Pour in enough vinegar to completely cover the egg and leave it for five minutes.

   What do you observe?

   _____

   What does this tell you?

   _____

3. Put the beaker and the egg where they will not be disturbed. You will leave the experiment there for five days.

   What do you observe after five days?

   _____

   _____

4. Carefully remove the egg using the spoon.

   Touch the egg shell carefully. What do you observe?

   _____

   _____

5. Clean the powder from the egg shell and rinse the egg in clean water.

   Describe the appearance of the egg now.

   _____

   _____

continued

Topic 3 Materials

6 Draw what your experiment looked like before and after the reaction took place.

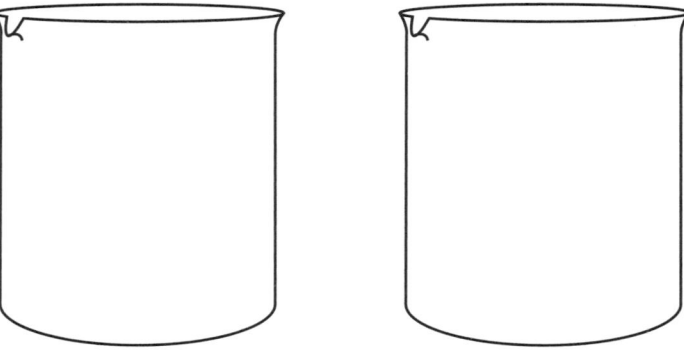

7 How could you tell that a chemical reaction had taken place?

_____

_____

8 Write a paragraph explaining what happened to the egg shell in this experiment.

_____

_____

_____

_____

_____

_____

_____

Topic **4** Forces and energy

Student's Book p **52**

4.1 Measuring force and drawing force diagrams

Force diagrams

Topic 4 Forces and energy

Student's Book p 52
4.1 Measuring force and drawing force diagrams

Make your own force meter

1 Follow these instructions to make a force meter.

Find a place to hang your force meter then set up your equipment like this:

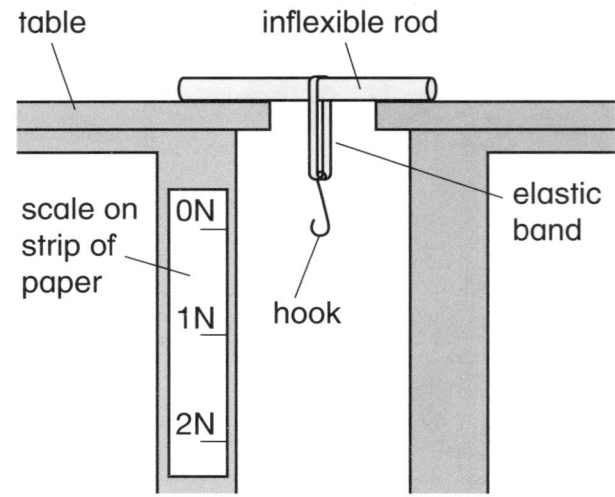

You will need:
- An inflexible wooden, plastic or metal rod
- A thick elastic band
- Two wire hooks (you can make these from large paper clips)
- A strip of paper
- A ruler
- Five average-sized apples or five 100 ml boxes of juice

The apple and the box of juice have a weight of about 1 newton (or 1 N). Use these to make a scale from 1 to 5 newtons (1 to 5 N) on the strip of paper.

Write a sentence explaining how your force meter works.

_____

_____

2 Find the weight of at least five different small items. Record your results in the table below.

Item					
Weight in newtons					

3 Use your results to draw a graph on some graph paper.

4 Discuss your data. How reliable is it? Do you need to repeat any measurements?

Topic 4 Forces and energy

Doubling and tripling the force

Student's Book p 52
4.1 Measuring force and drawing force diagrams

This is what some students tried. They wanted to make a force meter that could measure double the force of a force meter with a single elastic band.

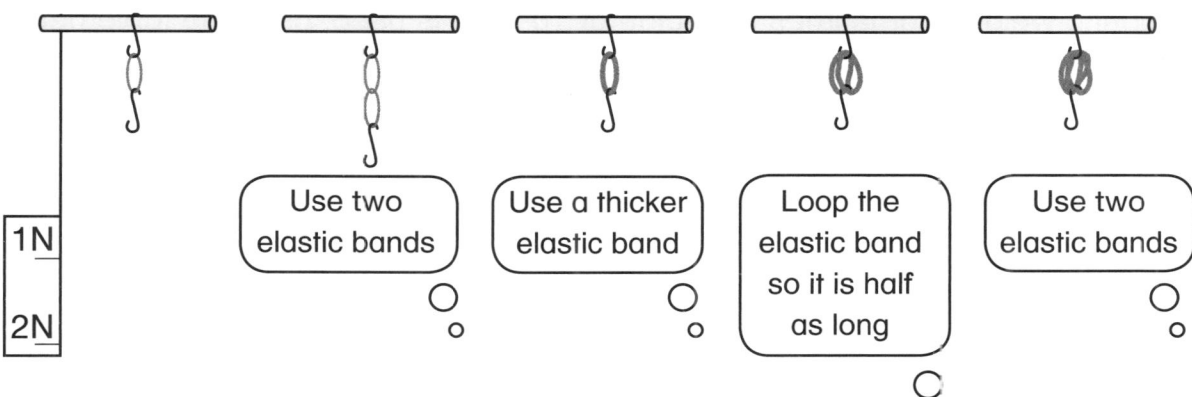

1 Which method do you think is best? Why?

   _____

   _____

2 Try the method you have chosen and comment on whether or not it is effective. Give reasons for your answer.

   _____

   _____

3 What will you do to make a force meter than can measure triple the force of a force meter with a single elastic band? Draw a labelled diagram to show your ideas.

4 Build and test your force meter. Comment on how well it worked or did not work. Do you need to repeat any measurements?

   _____

   _____

45

Topic 4 Forces and energy

Student's Book p 54
4.2 Weight, mass and gravity

Calculating the weight of different objects

The force of gravity on Earth is 9.8 N/kg, but to practise using the formula for calculating weight, sometimes the value of 10 N/kg is used, as in the table below.

1 Calculate the weight of the different objects on Earth. Fill in the table.
weight = mass × gravity in N/kg (newtons per kilogram)

Planet	Object	Mass (kg)	Gravity (N/kg)	Weight (N)
Earth	camel	400	10	4000
Earth	child	30	10	
Earth	cat	4	10	
Earth	elephant	5000	10	
Earth	tiger	300	10	

2 Calculate the weight of the different objects on the Moon. Fill in the table.

Planet	Object	Mass (kg)	Gravity (N/kg)	Weight (N)
Moon	camel	400	1.6	
Moon	child	30	1.6	
Moon	cat	4	1.6	
Moon	elephant	5000	1.6	
Moon	tiger	300	1.6	

Topic 4 Forces and energy

Student's Book p 54
4.2 Weight, mass and gravity

Calculating the weight on different planets

1 Calculate the weight of the following objects on different planets. Fill in the units on the tables.

Planet/ body	Object	Mass (____)	Gravity (N/kg)	Weight (____)
Mercury	child	30	3.78	111
Venus	child	30	8.9	
Earth	child	30	9.8	
Earth's Moon	baby elephant	130	1.6	
Earth's Moon	elephant	5000	1.6	
Mars	tiger	300	3.7	
Jupiter	cat	4	24.9	
Saturn	tiger	300	10.4	
Uranus	cat	4	8.9	
Neptune	child	30	11.2	

2 What can you say about these values of weight?

_____

_____

_____

47

Topic 4 Forces and energy

Student's Book p 54
4.2 Weight, mass and gravity

Investigate forces in different directions

1. What scientific question do you want to ask? _____

_____

2. Which of the five main types of scientific enquiry will you use to answer your question?

_____

3. Write your plan. Describe any risks and what you will do minimise them.

_____

_____

_____

4. What equipment will you need?

_____

_____

5. What will you measure and record? Identify the independent, dependent and control variables.

_____

_____

_____

6. Describe how you will compare your results.

_____

_____

_____

7. Carry out your investigation and record your results. Use a separate sheet of paper if you need more space.

8. What can you conclude about measuring forces from this experiment?

_____

_____

Topic 4 Forces and energy

Student's Book p 54
4.2 Weight, mass and gravity

Making sense of graphs

This graph shows the force that was needed (in newtons) to stretch five different rubber bands to twice their original length.

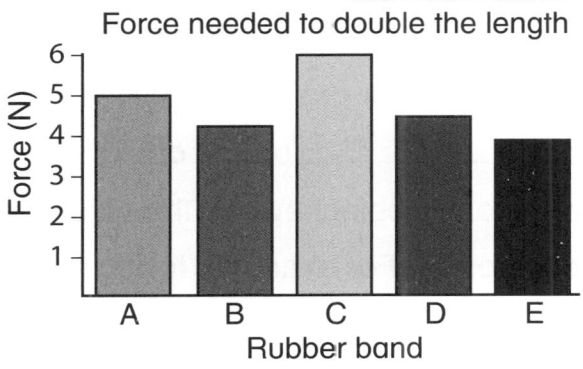
Force needed to double the length

1. Which rubber band needed the greatest force to stretch it to double its length?

   _____

2. Which rubber band is likely to have been thinnest? Why?

   _____

   _____

3. Which rubber band needed a force of 4.5 N to stretch it to double its length?

   _____

   Here is a different graph:

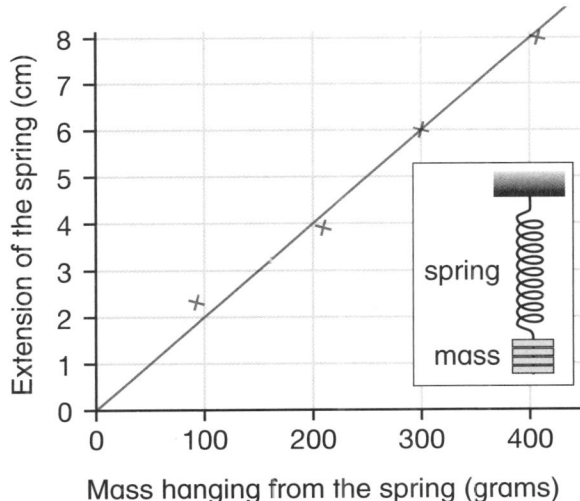

4. What does this graph show?

   _____

   _____

5. How much did the spring stretch when a mass of 300 g was hanging from it?

   _____

6. What is the weight in newtons required to stretch the spring by 4 cm?

   _____

   _____

Topic 4 Forces and energy

Forces and their effects

Student's Book p 56
4.3 The effect of different forces

1. Complete the sentences using words from the box.

    We can pull or push an object. When we do this, we are exerting a _____ on the object. A force from one object can act on another object even if the objects are not _____ each other. For example, falling objects are pulled to the ground because of _____.

    > gravity
    > touching
    > force

2. A ball on a snooker table travels fast in a straight line until it hits a cushion on the edge of the table. Then the ball bounces back off the cushion and slows down.

    Tick the **two best** sentences that describe what we can conclude from this.

    ☐ A force causes the ball to slow down.

    ☐ A force causes the total energy to decrease.

    ☐ A force causes the ball to change direction.

    ☐ Gravity causes the ball to slow down.

3. Look at the picture.

    a Draw a force diagram to show the force exerted by the girl and the door.

    b Explain why a closed door doesn't move when you push it like this.

Topic **4** Forces and energy

Student's Book p 58
4.4 Floating and sinking

Designing an experiment

1 What are you trying to find out? _____
   _____

2 What equipment will you need? _____
   _____

3 What will you do? _____
   _____

4 What will you do to make sure you are doing a fair test? Identify the independent, dependent and control variables.
   _____
   _____
   _____

5 What will you measure? _____
   _____
   _____

6 What will you do to make sure your results are reliable?
   _____
   _____

7 Describe how you will record your results so that it is easy to tell which boat performed best.
   _____
   _____

8 Carry out your test. Record your results on a separate sheet of paper.

9 Explain your results. _____
   _____
   _____

Topic 4 Forces and energy

Student's Book p 60
4.5 Staying safe in cars

Crash testing

Aim: *To model what happens to the human body in a car crash.*

Equipment: a ramp, a trolley with wheels, a brick or a block of wood, soft modelling clay, talcum powder or similar, sticky tape

Method

1. Make a model person using the clay. Coat the model with a layer of powder so it is not sticky.
2. Seat the model on the trolley.
3. Set up your ramp.

Test A

Let the trolley with the model person roll down the ramp onto the floor.

Record what happens to the person.

_____

_____

_____

Test B

Place the brick or wooden block on the floor close to the bottom of the ramp. Let the trolley with the model person roll down the ramp again.

Record what happens to the person this time. Suggest why.

_____

_____

_____

_____

What does the brick or wooden block represent in this test?

_____

continued

Topic 4　Forces and energy

Test C

Reshape your model person as necessary.

Use sticky tape to model a seatbelt for the model person.

Leave the brick or block in place and let the trolley roll down the ramp again.

What happens to the person this time? Suggest why.

_____
_____
_____
_____

If a passenger is injured in a head-on collision with another vehicle while not wearing a seatbelt, predict what sort of injuries they might have.

_____
_____
_____
_____

Topic 4 Forces and energy

Looking back

Student's Book p 62
Looking back Topic 4

Write a summary paragraph to explain each of the following: gravity, weight, effects of forces.

Gravity:

_____
_____
_____
_____
_____

Weight:

_____
_____
_____
_____
_____

Effects of forces:

_____
_____
_____
_____
_____

Topic **5** Light and electricity

Student's Book p **64**
5.1 Light

All about seeing

Draw arrows on the following diagrams to show how light allows the person to see the object. Then complete each sentence.

The light source in this diagram is the _____.

The boy can see the robot because light travels from the _____, reflects off the _____ and then travels to the boys eyes.

The light source in this picture is the _____.
_____ reflects off the water so that the boy can see the water.

55

Topic 5 Light and electricity

Student's Book p 64
5.1 Light

How we see things

1 Draw a picture of two light sources. Explain why we call these light sources.

_____   _____
_____   _____
_____   _____

2 Tick (✓) the correct boxes.

    a Can you see if there is no light?

       yes ☐   no ☐

    b Light from a light source _____ an object so that we can see the object.

       shines on ☐   reflects off ☐   blocks ☐

    c We can only see when light from a light source shines directly into our eyes.

       true ☐   false ☐

Topic **5** Light and electricity

Student's Book p **66**
5.2 Reflecting light

Building a periscope

You will need:
- a small cardboard box with a lid
- 2 small mirrors (or pieces of highly reflective material)
- scissors, sticky tape or glue

1. Measure the mirrors. Cut a hole roughly the same size as one mirror at one end of the box. Cut another hole, the same size, in the lid of the box.

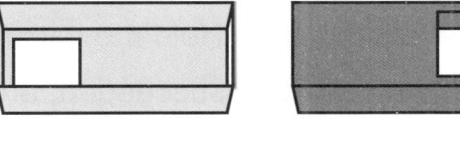

2. Put the mirrors in the box, each at an angle of 45°, at opposite ends of the box. The mirrors will be in front of the holes you have made.

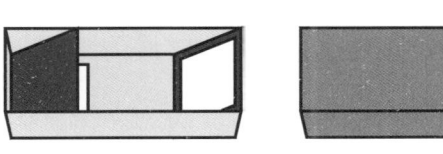

3. Glue or tape the mirrors in the correct positions inside the box.

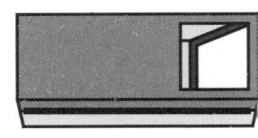

4. Put the lid on the box and tape it down. The holes must be at opposite ends of the box.

5. Explain briefly the way that your periscope works.

_____

_____

_____

_____

_____

57

Topic 5 Light and electricity

Student's Book p 66
5.2 Reflecting light

Can light move around a corner?

You will need:
- some mirrors or pieces of reflective paper
- a flashlight

1 Explore how you can get light to move around a corner. Use the mirrors to reflect the light. Hold the mirrors and the flashlight in different positions.

Once you have succeeded, draw a diagram in the box to show what you did.

2 Measure the angles of the mirrors and the distance between the light source and the mirrors. Add these measurements to your diagram.

3 Use your diagram to explain to the class the way that light can 'turn' around a corner.

Topic **5** Light and electricity

Student's Book p **68**
5.3 Bending light

Making a coin appear

You will need:
- a coin
- an opaque container such as a metal tin or a ceramic mug, and water

Method

1. Work with a partner. Put the coin into the container.
2. Look at the coin, then move further away until you can't see the coin. Stay where you are.
3. Ask your partner to gradually pour water into the beaker.
4. Tell your partner to stop pouring as soon as you can see the coin again.
5. Explain why you can see the coin again when there is water on top of it.

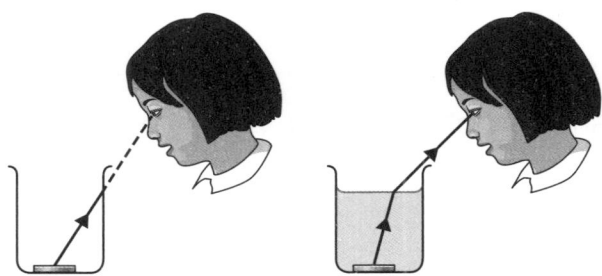

_____

_____

6. Can you make the coin appear to be deeper or shallower in the water by changing the water level? Investigate to find out.

   a. Describe what you did to answer the question.

   _____

   _____

   b. What can you conclude?

   _____

   _____

Topic 5 Light and electricity

Student's Book p 68
5.3 Bending light

Experiment with bending light

Our hypothesis: _____

1 Explain briefly what you will do to demonstrate this. Draw and label a picture to show the equipment you will use.

2 What do you predict will happen?

3 Draw diagrams to show what happened. Label your diagrams.

4 Write a few short sentences to explain what happened.

Topic 5 — Light and electricity

Student's Book p 70
5.4 Circuits and circuit diagrams

Using conventional symbols

Write the name of each circuit component below.

Draw a line to match each component to its correct symbol.

Name of component **Symbol**

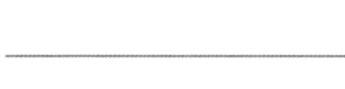 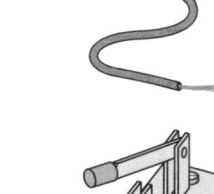 

 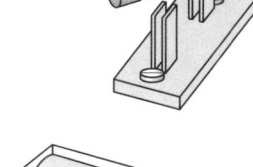 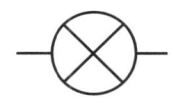

 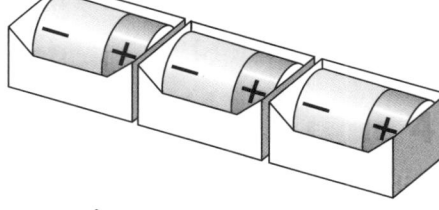 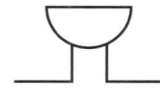

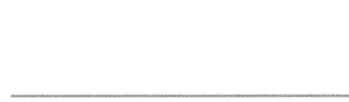 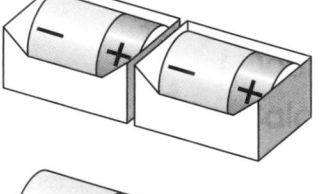 

 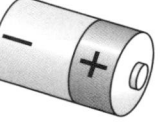 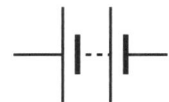

 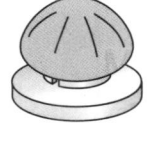 

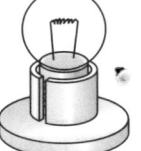 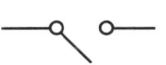

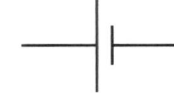

Topic 5 Light and electricity

Circuits and symbols

Student's Book p 70
5.4 Circuits and circuit diagrams

These circuit diagrams are all incorrect. Identify the errors in each diagram and redraw the circuit correctly.

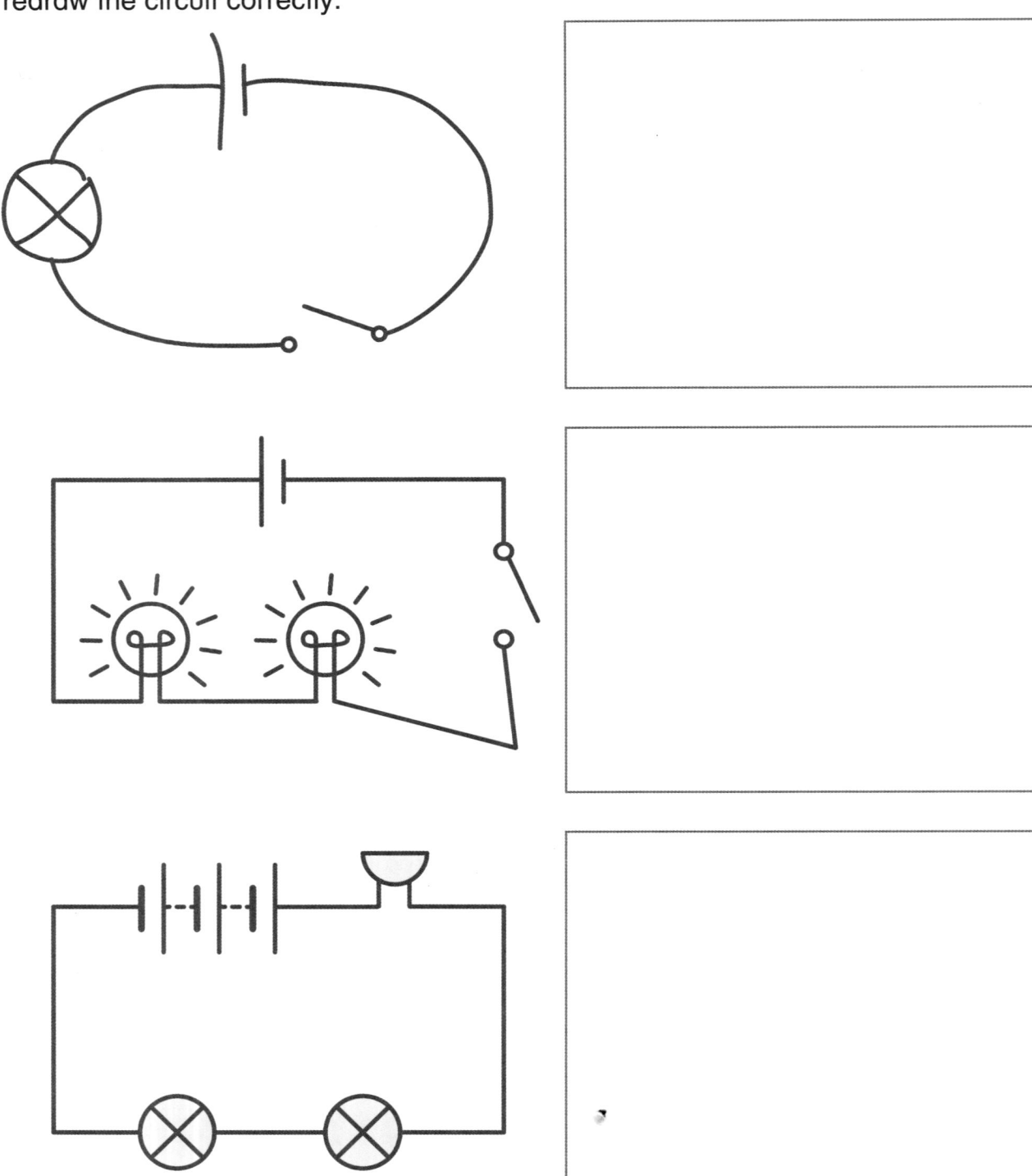

Topic **5** Light and electricity

Using diagrams to build circuits

Student's Book p 70
5.4 Circuits and circuit diagrams

1 What does each of these symbols represent?

a _____ b _____ c _____ d _____

2 Make these circuits one by one. Close the switch to test it. Check with your teacher that each circuit is correct.

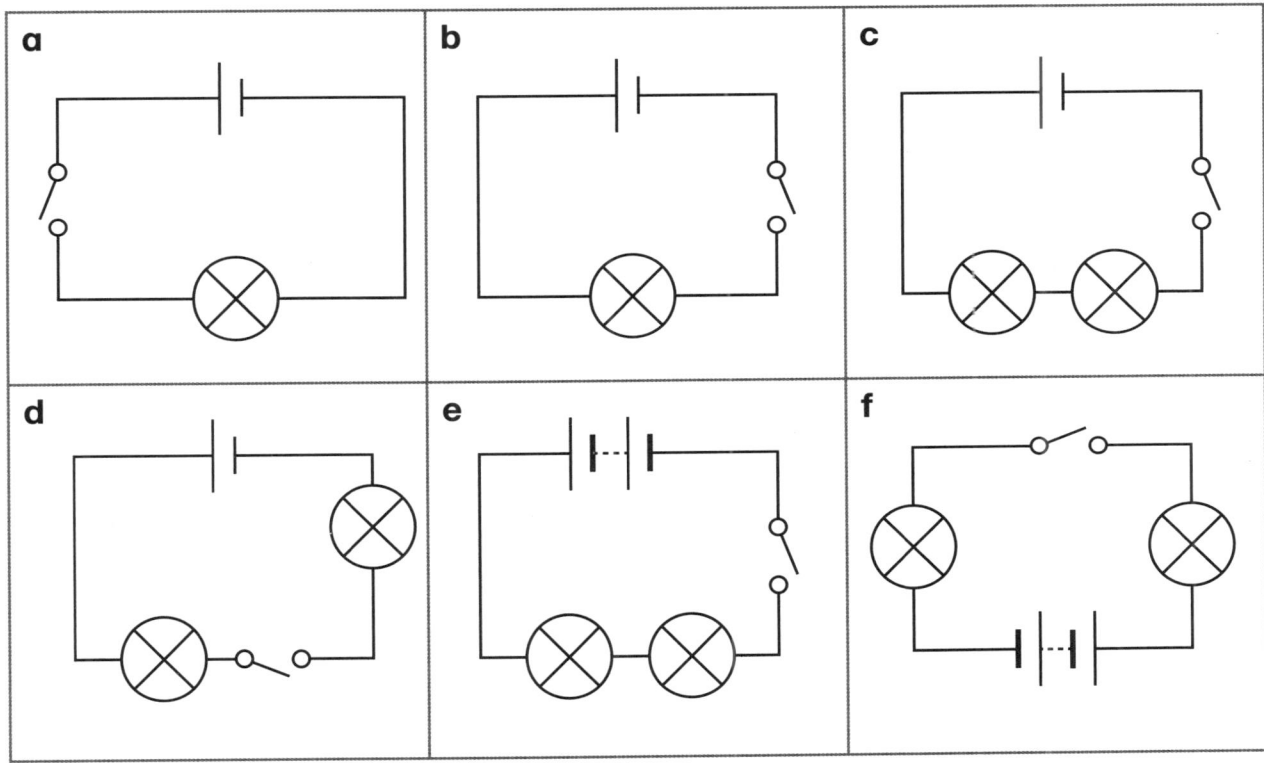

3 What happens if you unscrew one lamp in a circuit with two lamps? Why does this happen?

_____

_____

_____

63

Topic 5 Light and electricity

Student's Book p 72
5.5 Series circuits

Testing series circuits

Make and test each of these series circuits. Draw the circuit diagrams in the boxes next to each circuit and say how bright the lamp or lamps are.

Circuit	Circuit diagram	How bright are the lamps when you close the switch?

Topic **5** — **Light and electricity**

Student's Book p **72**
5.5 Series circuits

Changing components

Does it matter whether you add a buzzer or an extra lamp to a circuit with one lamp? Plan a fair test investigation to identify how changing the type of component affects the brightness of the lamp.

1 What are you trying to find out?

_____

_____

2 Describe how you will do this.

_____

_____

_____

_____

3 What equipment will you need?

_____

_____

4 What will you do to make sure you are doing a fair test? Identify the independent, dependent and control variables.

_____

_____

_____

_____

5 Carry out your test. Record your observations on a separate sheet of paper.

6 Do a buzzer and an extra lamp have the same effect on a circuit? Explain why.

_____

_____

_____

Topic 5 Light and electricity

Student's Book p 74
5.6 Parallel circuits

Comparing circuits

Aim: *To construct and compare a series and a parallel circuit*

Equipment: two identical lamps, connecting wires, one cell

Method:

1 Build the circuit shown in Circuit A. Observe the brightness of the single lamp.

Circuit A

Circuit B

2 Add a lamp in series to Circuit A. Observe the brightness of each lamp now.

   a Draw a circuit diagram of your modified circuit (Circuit B).

   b What is the effect of adding a lamp?

_____

3 Modify the circuit so that the lamps are connected in parallel. Compare the brightness of the lamps again.

Draw a circuit diagram to show your new circuit (Circuit C).

4 How does the brightness of two lamps in series and two lamps in parallel compare with the brightness of a single lamp?

Circuit C

_____

_____

5 Predict the effect of adding a further lamp in series to Circuit B.

_____

6 Predict the effect of adding a further lamp in parallel to Circuit C.

_____

Topic **6** Rocks and soil

Student's Book p **78**
6.1 Measuring and calculating

Calculate percentages and graph them

Use the measurements given on the diagrams to work out the percentage of sand, silt and clay in each sample.

Use the blank pie chart to show the proportions of sand, silt and clay in each sample.

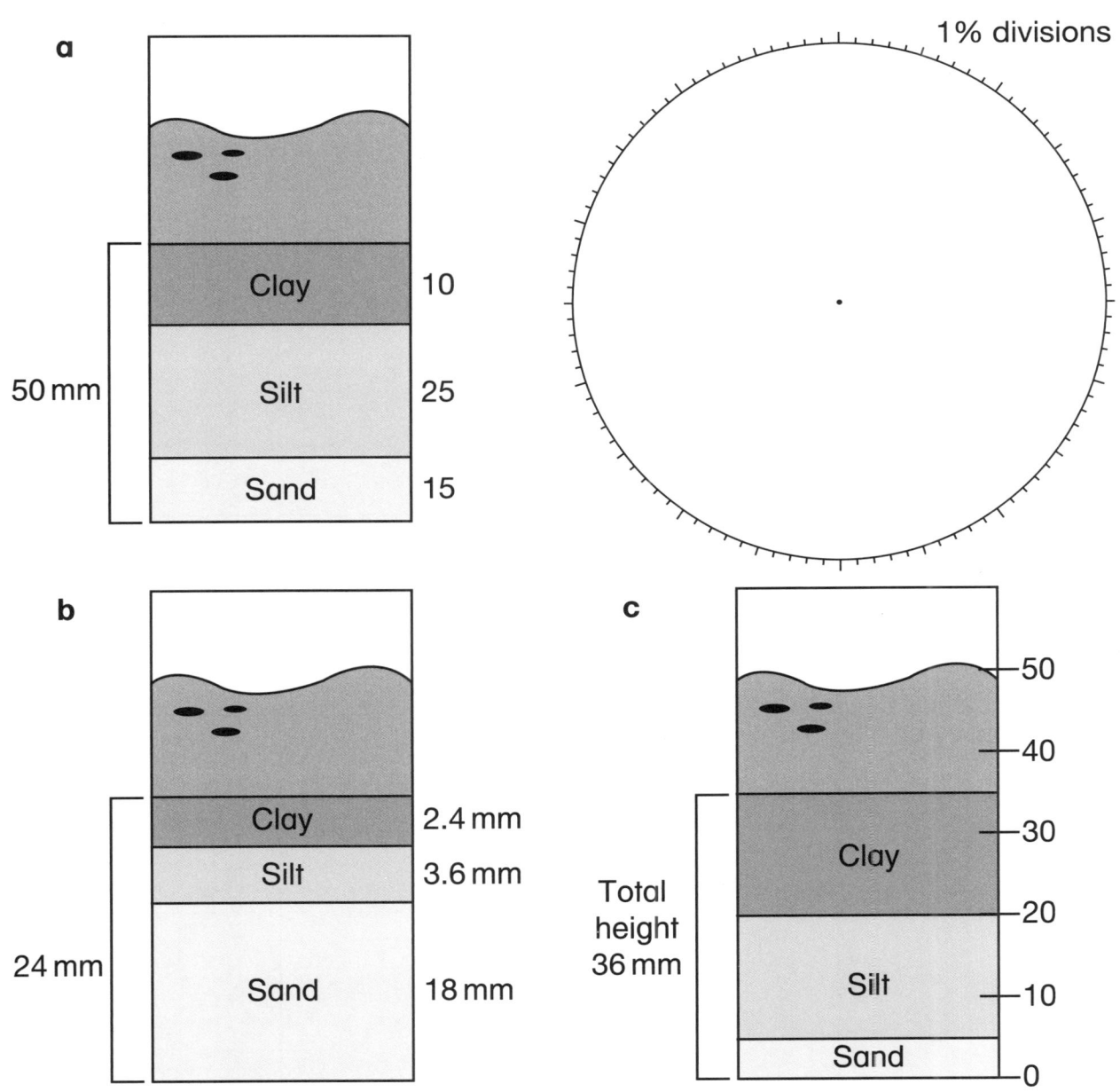

67

Topic 6 Rocks and soil

Observing rock samples

Student's Book p 80
6.2 Rocks

Carefully observe the three rock samples you collected. Complete this table to summarise your observations. Draw and label what you see under the magnifying glass.

Characteristics	Sample 1	Sample 2	Sample 3
Overall appearance			
Colour			
Shape			
Texture (rough or smooth)			
Mass (heavy or light)			
Other			

Topic 6 Rocks and soil

Student's Book p 80
6.2 Rocks

Classifying rocks using a key

1 Do you remember how to make and use a branching key? Read the information carefully before you start.
For example:

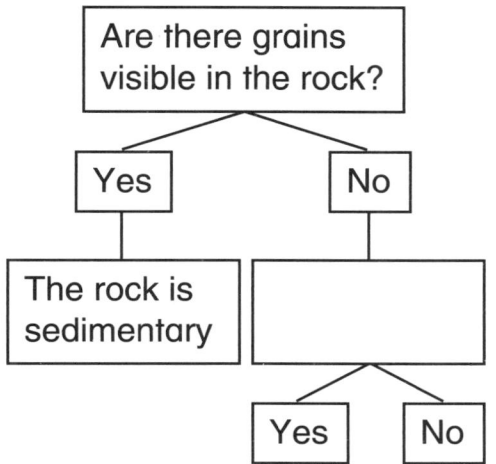

- A branching key is used to identify things by asking questions about their characteristics.
- At each step the key asks a question and gives two choices for the answer (usually 'yes' or 'no').
- Each choice leads to another question until the object is identified.
- The key can be drawn like a branching tree diagram.

2 Make a simple branching key to help someone decide whether a rock is igneous, sedimentary or metamorphic. Add steps to the key as you need to.

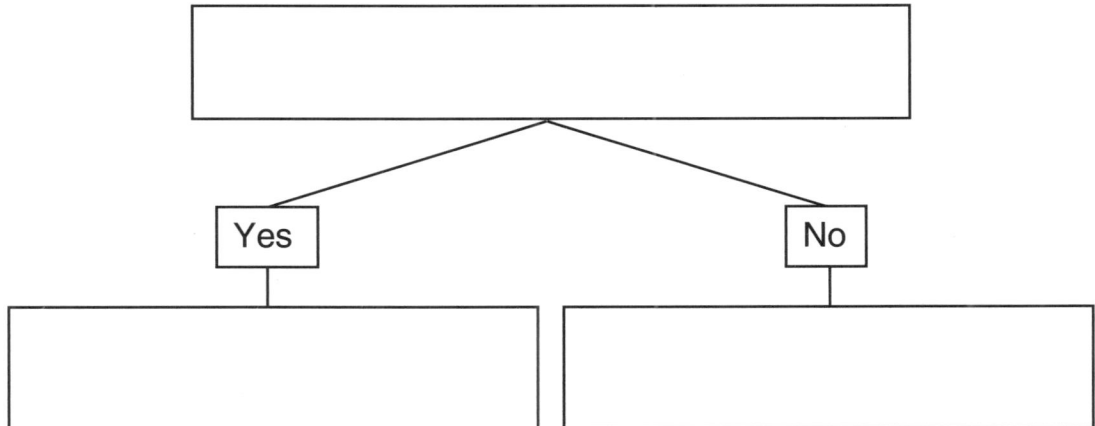

Topic 6 Rocks and soil

Student's Book p 82
6.3 The rock cycle

The rock cycle

This is a diagram showing the rock cycle. Complete the diagram by filling in the processes that take place at each stage.

The rock cycle

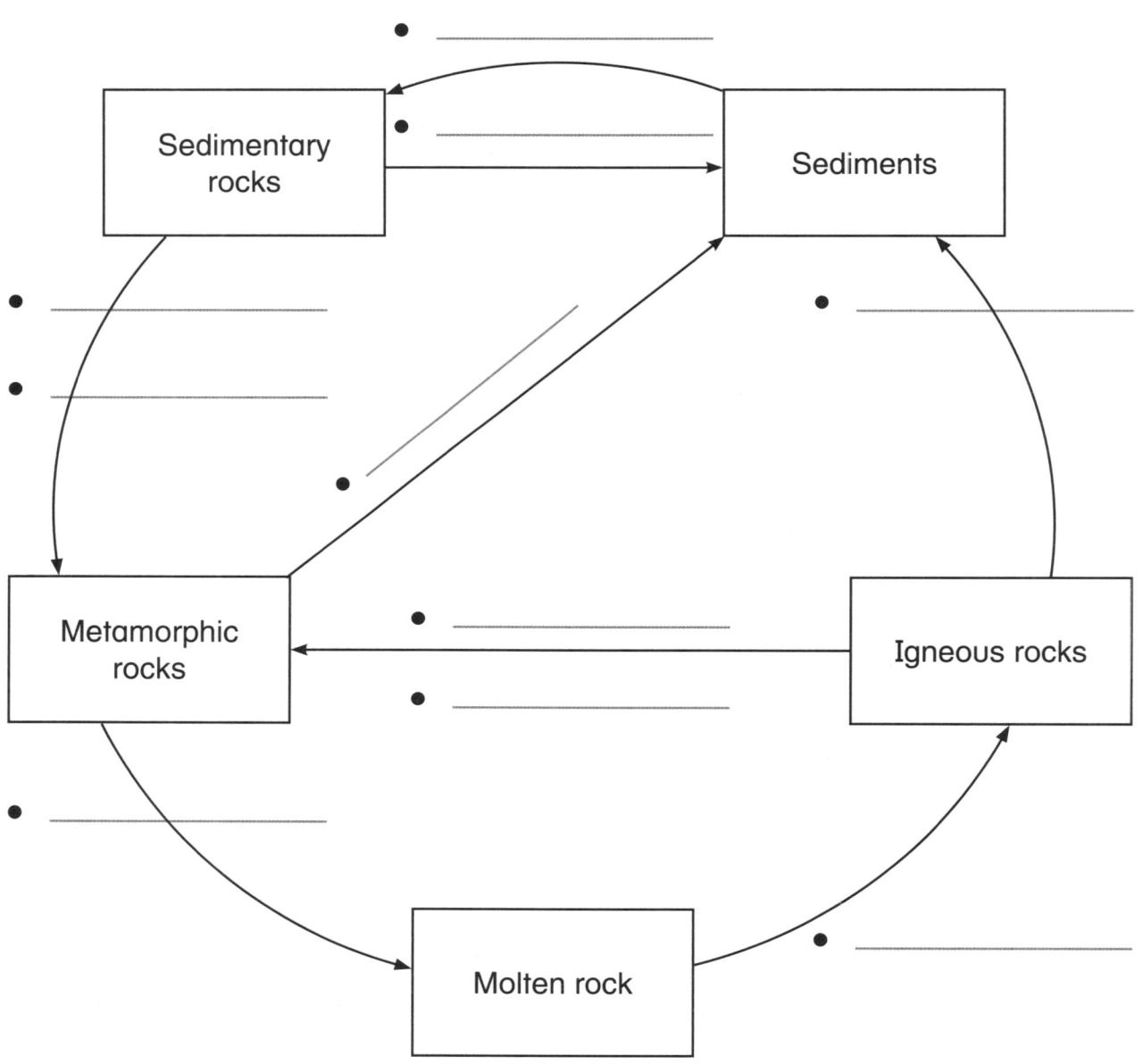

70

Topic **6** Rocks and soil

Student's Book p 84
6.4 Fossils

Observe and compare your fossils

1 Look closely at your fossils and then complete the table below.

Observations	Fossil 1: Cast	Fossil 2: Mould
Hair		
Shape		
Texture		
Size		
Colour		
Other		

2 Which fossil gives you the most information, the mould or the cast (3D)?

_____

3 Why?

_____

_____

Topic 6 Rocks and soil

Student's Book p 86
6.5 Soil

Which type of soil holds water best?

This is the question we are asking:

_____

_____

Independent variable (this is what we will change):

_____

_____

_____

Control variable (these are the things we will keep the same):

_____

_____

_____

Dependent variable (this is what we are going to measure):

_____

_____

_____

This is a list of the equipment we will need to do our test:

_____ _____

_____ _____

_____ _____

_____ _____

continued →

Topic 6 Rocks and soil

Diagram showing our experiment set up:

Our observations:

Our conclusion:

Topic 6 Rocks and soil

Which type of soil?

Student's Book p 88
6.6 More about soil

1 The same amount of water was poured into samples of clay, loam and sandy soil. The diagram shows the results.

   a Which funnel contains the loam sample? Give a reason for your answer.

   _____

   b Which sample do you think drained the fastest? Why?

   _____

   c Identify the sandy soil and the clay soil based on these results.

   _____

 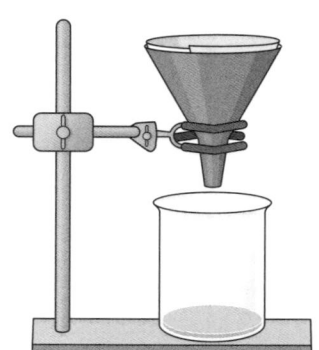 

2 An article about soil improvement contains the following information.

| When soil is depleted of organic material, it is less absorbent and water tends to settle on the surface and run off. | Increasing the organic content of soil by just 1% results in soil retaining 75 000 litres of water per acre. (An acre is about $\frac{3}{5}$ the size of a soccer pitch.) |

continued

a How does organic material get depleted in soils?

_____

b How can gardeners and farmers increase the organic content of soil?

_____

_____

3 Suggest at least two advantages of increasing the water-holding capacity of soil.

_____

_____

Topic 7 Earth in Space

Modelling the orbits of different planets

Student's Book p 92
7.1 The Solar System

> **You will need:**
> - a long string with a rubber bung attached
> - a short string with a rubber bung attached

1. Whirl the bung with a short string in a circle. Watch the way the bung moves.
2. Now try whirling the bung with the long string.
3. Repeat with each string.
4. Draw a picture of the investigation.

continued

5 Fill in each of the gaps with *shorter* or *longer*.

The bung that needed more effort to whirl it had the _____ string.

The bung that needed less effort to whirl it had the _____ string.

6 Explain why you repeated the activity with each bung.

_____

_____

_____

Imagine that, in your investigation, you were comparing the orbits of Earth and Neptune.

7 Which bung represented Earth? _____

8 Which bung represented Neptune? _____

9 Name a planet that you could represent by a bung with a medium-length string.

_____

Topic 7 Earth in Space

Student's Book p 92
7.1 The Solar System

Comparing how fast the planets move

The planets in the Solar System orbit the Sun. Some move faster than others.

1 Complete this table using data from the table on page 93 of your Student's Book.

2 Draw a bar chart to show how fast each of the planets move.

Planet	How fast it moves (km/second)
Mercury	
Venus	
Earth	
Mars	
Jupiter	
Saturn	
Uranus	
Neptune	

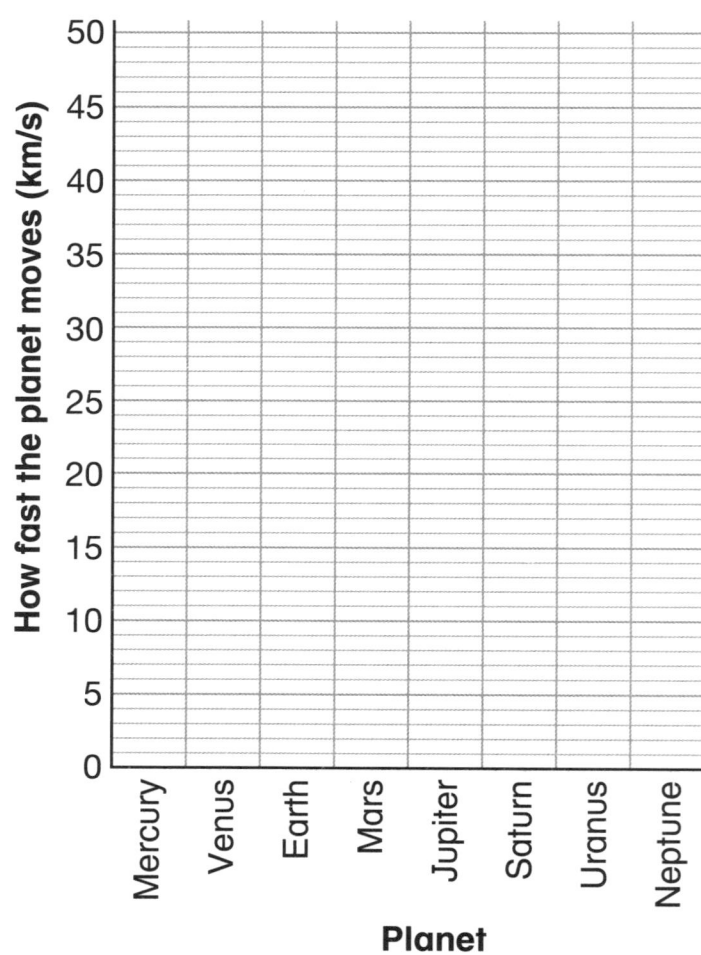

3 Think about the order of the planets from the Sun. Which one is the closest? Which one is furthest away? Can you see any link between the distance from the Sun and how fast the planet moves?

Topic 7 Earth in Space

Student's Book p 92
7.1 The Solar System

Pluto – a planet no more

1. Read this newspaper article carefully.
2. Re-read the article and highlight the main points.

Astronomers have voted to strip Pluto of its status as a planet.

— Astronomy News, September 2006

About 2500 scientists meeting in Prague have voted for Pluto to be reclassified as a dwarf planet.

The researchers said Pluto failed to dominate its orbit around the Sun in the same way as the other planets.

Pluto was discovered in 1930 by the American Clyde Tombaugh.

The decision is likely to upset the public, who have become accustomed to nine planets in the Solar System.

Teary-eyed

"I have a slight tear in my eye today, yes; but at the end of the day we have to describe the Solar System as it really is, not as we would like it to be," said Professor Iwan Williams.

The need for a strict definition for a planet was seen as necessary after new telescope technologies began to reveal far-off objects that were similar to Pluto in size. Without new rules, textbooks could soon be talking about 50 or more planets in the Solar System.

The scientists agreed that for a body to qualify as a planet:

- it must be in orbit around the Sun
- it must be large enough that it is almost spherical
- it has cleared its orbit of other objects.

Pluto was automatically disqualified because its elongated orbit overlaps with that of Neptune.

Pluto's status has been debated for many years. It is further away and considerably smaller than the eight other planets in our Solar System. Pluto is smaller than some moons in the Solar System, including our own Moon. Its orbit around the Sun is also tilted compared with the plane of the other planets.

In addition, since the 1990s, astronomers have found several objects of comparable size to Pluto in outer regions of the Solar System. The final blow for Pluto came with the announcement in 2005 of the discovery of an object now called Eris. After being measured with the Hubble Space Telescope, it was shown to be bigger than Pluto.

continued

Topic 7 Earth in Space

3 What is Pluto now known as?

_____

4 What are the rules for being classified as a planet?

_____

_____

_____

5 Why did Pluto not remain as a planet?

_____

_____

6 How have dwarf planets been seen?

_____

_____

7 Why do you think some scientists have been upset by the decision to classify Pluto as a dwarf planet?

_____

_____

8 What is the current thinking about the status of Pluto? Do some reading of your own to find out.

_____

_____

_____

Topic 7 Earth in Space

Student's Book p 94
7.2 The Moon

Observing the Moon

Use the table below to record your observations about the Moon. If you could not see the Moon clearly, circle the observation to show that you have guessed.

Date	Observation	Date	Observation

Topic 7 Earth in Space

Phases of the Moon

Student's Book p 94
7.2 The Moon

The diagram shows the positions of the Sun, Earth and Moon at eight phases in the lunar month. Copy the labels and images from the box into the correct places in the diagram.

new moon
full moon
half moon (third quarter)
waxing crescent
waning gibbous
half moon (girst quarter)
waxing gibbous
waning crescent

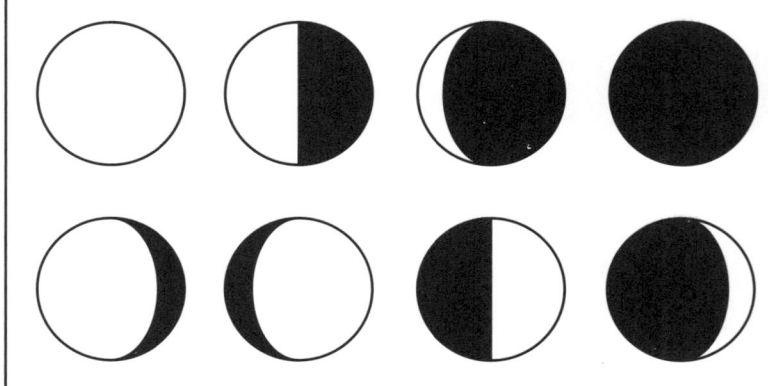

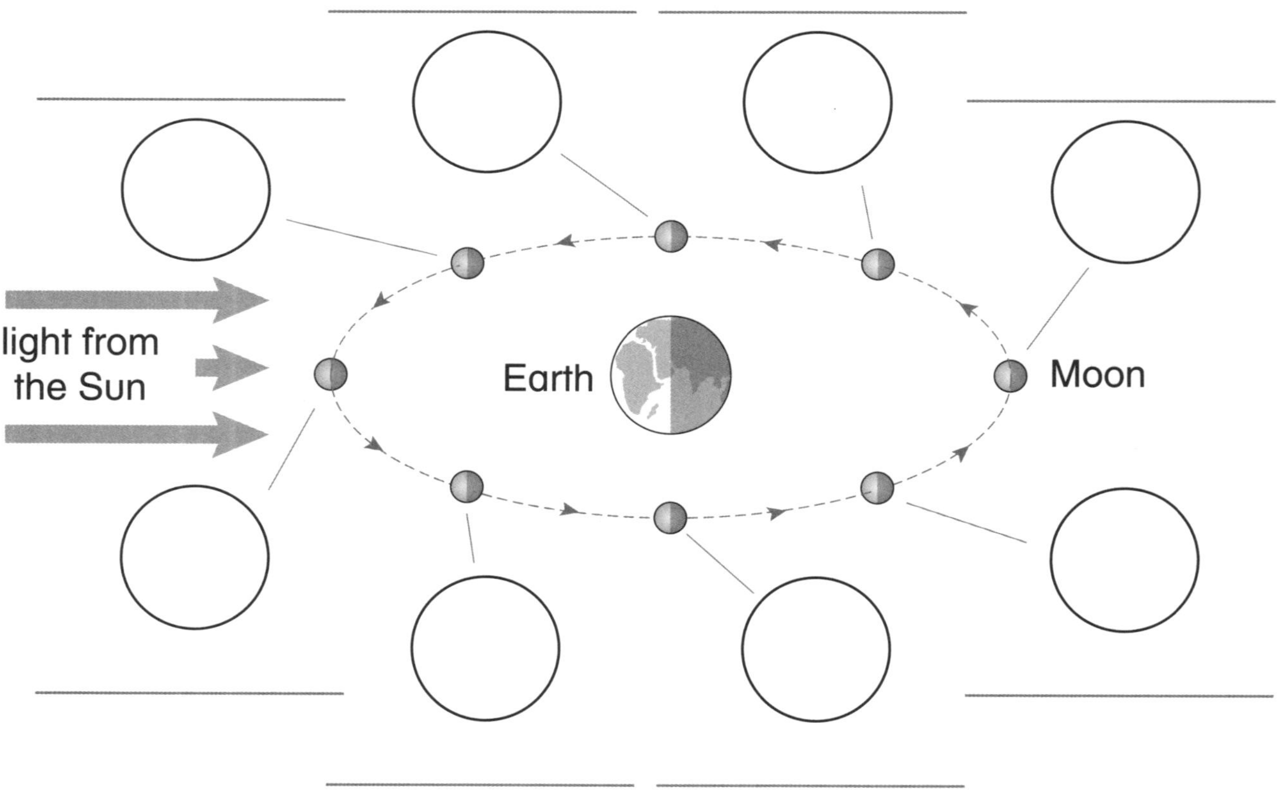

Appendix 1

Electrical symbols

Cell	
Battery of cells	
Wire	
Lamp	
Junction of conductors	
Open Switch	
Closed Switch	
Buzzer	

Units for physical quantities

Quantity	Units
Length (used for length, height and width)	mm, cm, m, km
Area	cm^2, m^2
Volume	ml, l, cm^3, m^3
Weight	N
Mass	g, kg
Time	s, min, h
Force	N
Gravity	N/kg
Temperature	°C